PROVEN PROFITS FROM POLLUTION PREVENTION:
CASE STUDIES IN RESOURCE CONSERVATION AND WASTE REDUCTION

Donald Huisingh, Ph.D.
Larry Martin
Helene Hilger
Neil Seldman, Ph.D.

Institute for Local Self-Reliance
2425 18th Street, NW
Washington, D.C. 20009
(202) 232-4108

Since 1974, the Institute for Local Self-Reliance (ILSR), a research and educational organization, has provided technical information and assistance to city, county, and state governments, and citizen and neighborhood organizations. ILSR focuses on energy and waste utilization from a common sense economic development perspective. ILSR works exclusively in urban areas. Current ILSR projects are located in Philadelphia, Pennsylvania; Saint Paul, Minnesota; Newark, Trenton, and Edison, New Jersey; Cleveland, Ohio; Fayetteville, Charlotte, Durham, and Asheville, North Carolina; Providence, Rhode Island; Boston, Cape Cod, and New Bedford, Massachusetts; New Haven, Connecticut; Washington, D.C.; and Gainesville, Florida.

Staff of the Institute for Local Self-Reliance:

Co-directors: David Morris, Ph.D.
              Neil Seldman, Ph.D.

Energy and Economic Development

David Morris, Ph.D.
John Plunkett
Jack Gleason, M.S.
David Peterson, L.L.B., M.A.

Waste Utilization/Economic Development

Neil Seldman, Ph.D.
Larry Martin
Ken Woodruff, M.E.
Rick Anthony, M.A.
Don Huisingh, Ph.D.
Mary Appelhof, M.S.
Donald Rowe

ROBERT
STRO

Administration

SEP 12

David Bardaglio
Jan Simpson

Tallahassee, Florida

Cover design by James True
Printed in the United States of America

CONTENTS

Acknowledgements   i

Foreword ii

Preface iii

CHAPTER 1:   INTRODUCTION TO HAZARDOUS
             WASTE MANAGEMENT ISSUES   1

Historical Perspective   1
Types and Quantities of Hazardous Waste Generated   4
The Cost of Hazardous Waste Generation to
   Industry, Society, and the Environment   9
Pollution Prevention-Waste Minimization:
   Essential for Economic and Human Health 13

CHAPTER 2:   POLLUTION PREVENTION-WASTE
             MINIMIZATION CONCEPTS AND APPROACHES 16

Overview of Pollution Prevention and Waste Minimization . . 16
Review of Individual Components and Processes Applicable
   for Pollution Prevention and Waste Minimization 17

CHAPTER 3:   WASTES TO RESOURCES 24

Characterization of Wastes Not Produced, Recovered, or
   Recycled as a Consequence of Strategies Documented
   in the Case Studies 24
Waste Reduction Projections 26

CHAPTER 4:   CASE STUDIES ORGANIZED BY
             STANDARD INDUSTRIAL CODE 27

Agriculture Forestry & Fisheries   0100 & 0200

Case Study 1: The Choplin Farm 28

Mining and Extractive Industries   1000

Case Study 2: Climax Molybdenum Company 33

Food And Kindred Products   2000

Case Study 3: California Canners and Growers 37
Case Study 4: Dunbar Foods 41
Case Study 5: Miller Brewing Company 45

Textile Mill Products   2200

Case Study 6: Hampshire Hosiery 49

Furniture   2500

Case Study 7:   United Globe Corporation 53

Paper and Allied Products   2600

Case Study 8:   West Point Pepperell 58

Printing and Publishing   2700

Case Study 9:   Rexham Corporation 62

Chemicals and Allied Products   2800

Case Study 10:   Exxon Chemicals America 67
Case Study 11:   Allied Chemical 71
Case Study 12:   Borden Chemical Company 74
Case Study 13:   American Enka Company 79
Case Study 14:   Minnesota Mining and Manufacturing (3M) . . 82
Case Study 15:   The Coty Division of Pfizer, Inc. 85
Case Study 16:   USS Chemicals 88
Case Study 17:   Industrial and Agricultural Chemicals, Inc. 92
Case Study 18:   ICI Americas, Inc. 95
Case Study 19:   Daly-Herring Company 99
Case Study 20:   Texasgulf 102

Rubber Plastic Products   3000

Case Study 21:   Nebraska Small Farms Action Group 105

Metals   3300

Case Study 22:   Florida Steel Corporation 109
Case Study 23:   Waupaca Foundry 112

Fabricated Metal Products   3400

Case Study 24:   Stanadyne, Inc., Moen Group 117
Case Study 25:   Elkhart Products Division, Inc. 130
Case Study 26:   Pioneer Metal Finishing, Inc. 133

Non-Electric Machinery   3500

Case Study 27:   Deere and Company 136
Case Study 28:   Emerson Electric Company (Also 3600) 141

Electrical and Electronic Equipment   3600

Case Study 29:   Hamilton Beach Division of Scovill, Inc. . . 149
Case Study 30:   GTE Sylvania 154
Case Study 31:   Data General Corporation 158
Case Study 32:   Minnesota Mining and Manufacturing (3M) . . 165
Case Study 33:   Digital Equipment Corporation 168

Transport Equipment   3700

Case Study 34:   Modine Manufacturing 171

Instruments and Related Products   3800

Case Study 35:   Rexham Corporation 175

Electric, Gas & Water   4900

    Case Study 36: Carolina Power and Light Company 182
    Case Study 37: Duke Power Company 186
    Case Study 38: Natural Power 196
    Case Study 39: Hickory/Newton/Conover Complex 200
    Case Study 40: City of Morganton, North Carolina 204
    Case Study 41: San Diego Regional Water Reclamation
                    Agency 213

Wholesale Trade, Non-Durable Goods   5100

    Case Study 42: Chevron USA, Inc. 221

Service Industries   7000

    Case Study 43: Minnesota Mining and Manufacturing (3M) . . 226
    Case Study 44: PCA International, Inc. 229
    Case Study 45: American Fotokemi, Inc. 235
    Case Study 46: Vulcan Automotive Equipment Ltd. 239

    CHAPTER 5:   ABSTRACTS CROSS-REFERENCING CASE
                 STUDIES BY MATERIALS 243

Agriculture/Food Processing Wastes 244
Fly Ash and Bottom Ash 245
Low-Level Radioactive Wastes 246
Metals and Other Inorganic Chemicals 247
Municipal Wastes 254
Process Water 256
Rubber and Plastics 260
Solvents, Oils, and Other Organic Chemicals 261

    CHAPTER 6:   ABSTRACTS CROSS-REFERENCING CASE
                 STUDIES BY PROCESS 268

Adsorption 269
Distillation 270
Electrolytic Reactions 272
Equipment Design and Operation 275
Filtration 282
Housekeeping 283
Incineration/Heat-Energy Recovery 285
Ion Exchange 287
Land Application 289
Management Initiatives 290
Materials Substitution 293
Recycling, Recovery, and Reuse 295

    CHAPTER 7:   INCREASING THE RATE AT WHICH INDUSTRIES AND
                 INSTITUTIONS IMPLEMENT POLLUTION PREVENTION AND
                 WASTE MINIMIZATION POLICIES AND TECHNOLOGIES . . 304

Goals and Objectives for Waste Minimization
    Throughout Society 304
A Discussion of Impediments to the Implementation
    of Pollution Prevention-Waste Minimization
    (PP-WM) Approaches 305

Proposed Strategy for Overcoming the Barriers and
    Accelerating the Global Implementation of
    PP-WM Approaches 308

CONCLUSION 311

LITERATURE CITED 312

APPENDIX 1: The Questionnaire Used in Gathering Data . . . 313

ACKNOWLEDGEMENTS

The authors gratefully acknowledge the cooperation of industrial leaders who have provided information about the ways they have reduced the quantities of hazardous and non-hazardous waste they produce, while simultaneously increasing net profitability.

Our thanks to the North Carolina Board of Science and Technology for a grant to perform the first portion of this study and for their permission to utilize some of the North Carolina case study materials in this present volume. Also, our grateful appreciation to Sven Thesen who contributed much to the first edition.

Our thanks to the Institute for Local Self-Reliance for providing timely assistance to the project.

Our thanks to our families who, with understanding and patience, have encouraged us to devote the time and effort necessary to develop this volume in the very short time allotted.

Our gratitude also to all who build upon the examples included here, and thereby help to make the world a safer and more healthy place to live.

FOREWORD

Since 1974, the Institute for Local Self-Reliance has worked to make communities more efficient, and to increase the amount of work derived from local resources. Higher efficiency makes for a healthier local economy. It also makes communities more self-reliant.

The concept of local self-reliance has broad application. Rather than import all our materials from outside sources and dump the used materials in waste repositories, the self-reliant community embraces closed-loop systems that minimize the use of materials and maximize recycling. Rather than base the health of a local economy on its ability to attract jobs from remote corporations, the self-reliant community generates employment by nurturing its native entrepreneurialism and inventiveness. Rather than allow remote governments to make decisions that destabilize the local economy, the self-reliant community assumes the authority and responsibility for its own future.

The Institute works with interested communities and provides technical information to the general public on innovative, locally based solutions to the problems facing our society. Proven Profits from Pollution Prevention is one in a series of such books. Twenty years ago, Americans discovered that some of the materials we thoughtlessly consumed and discarded were threatening our health. The realization led to a series of environmental laws intended to clean up the air, water, and land. By imposing the social costs of pollution on manufacturers, the nation significantly changed the dynamics of industrial waste disposal. A new technological dynamic emerged. Manufacturers learned to produce the same products and services with greater efficiency and far less pollution.

This book is a testament to the ability of society to channel human ingenuity. By setting environmentally and socially sensitive design parameters for engineers, we encourage ways of doing things that are supportive of the healthy progress of society as a whole. Proven Profits from Pollution Prevention presents the case and demonstrates that efficiency at the local level must be a major remedy for our toxic epidemic.

David Morris, Co-director
Institute For Local Self-Reliance
July 1985

ii

PREFACE

This volume had its conceptual origin in May of 1982 at the close of North Carolina's first "Pollution Prevention Pays" conference. Dr. Michael Royston of Geneva, Switzerland suggested that industry and government would soon be needing an easy-to-use compendium of case studies illustrating how various industries have captured the essence of the source reduction concept by meeting environmental regulation's goals through primary emphasis on pollution prevention instead of on pollution control.

In the following months, enthusiasm for the pollution prevention concept grew among North Carolina industrialists, governmental regulators, educators, and environmentalists. This interest was translated into a financial commitment to the pollution prevention concept when the legislature allocated $300,000 to the North Carolina Board of Science and Technology to sponsor a wide array of educational and research activities designed to help more industrial leaders become acquainted with the approaches and benefits of pollution prevention procedures.

Don Huisingh and Helene Hilger applied for and received a small grant to develop both the methodology for obtaining information from industries, and for presenting it in a readily usable form. A copy of the questionnaire used in the data gathering phase is included in Appendix 1. A first volume of materials derived from these information-gathering efforts was published through the North Carolina Board of Science and Technology in May 1985 under the title: A Compendium of North Carolina Case Studies in Resource Conservation and Waste Reduction: Profits of Pollution Prevention. That volume included information drawn exclusively from North Carolina-based industries. With permission from the Board of Science and Technology, we have included some of those North Carolina case studies, and have added others prepared from replies to our questionnaire from industrialists in other states and from a Canadian respondent.

The present volume has been made possible by the Institute for Local Self Reliance. The Institute's Waste Utilization staff and resources made possible the extension of the scope of research and guaranteed widespread and immediate dissemination of this information. Our thanks to the agencies who supported the Institute's research in hazardous waste minimization strategies and made this publication possible. These agencies are: the Beldon Fund, Washington, D.C.; the Ruth Mott Fund, Flint, MI; and the Arca Foundation, Washington, D.C.

CHAPTER 1

Introduction to
Hazardous Waste Management Issues

HISTORICAL PERSPECTIVE

Humans have long been aware of the dual nature of substances. On the one hand, they provide benefits, on the other, they present risks. The esthetically beautiful lead ceramic glazes used on table china are thought to have contributed to the demise of the Roman Empire due to the lead which leached from them. Felt hats were attractive, but also resulted in the "mad hatter" neurological syndrome in 19th-century hatmakers who were exposed to mercury in the hatmaking process. Many other similar examples could be cited, but for present purposes, our focus will be the dramatic increase in societal reliance on new synthetic substances that has occurred since World War II.

People were anxious to chart a new course after the war and were ready to enjoy the "better life through chemistry," as one leading company advertised. They demanded an increasing supply of consumer products. Industry responded by producing numerous products that utilized thousands of new compounds. Billions of metric tons of these new materials were soon produced and put into use.

The western style of "use it once and throw it away" has become normative; so much so that whole armies of workers are employed in picking up and hauling "away" our "wastes."

Confidence in science and technology was shaken in 1962 when Rachel Carson's book, Silent Spring, alerted the public to the fact that DDT, the symbol of human triumph over the insect world, could be found in organisms everywhere on earth, thus underscoring the global interconnection of ecosystems. She predicted that DDT would bring about the extinction of a number of species and might even threaten the human species. While these predictions were debated by scientists and politicians on many fronts, the public began to question the consequences and implications of a "better life through chemistry."

During the 1970s and '80s, nations around the world became aware of the multiple impacts of industrialized societies on human health, air quality, water quality, land quality, and food quality. They demanded changes. Consequently, many pieces of pollution-control legislation were enacted. The laws were enacted on a piecemeal basis, focusing, in turn, upon air pollutants, water pollutants, workplace pollutants, and food pollutants. The laws were frequently enforced by separate divisions of separate departments within state and federal governments, resulting in compartmentalization and fragmentation in environmental protection efforts. Consequently, air pollutants were "controlled" and kept from entering the air, but often the pollution-control sludges were then dumped into the water. Or, pollutants were kept from entering the water but were placed on or in the land. To our chagrin, these substances then moved from the pits, ponds, lagoons, and landfills, into surface water and groundwater, and, from there, into the global food chain!

The public became increasingly impatient with and distrustful of politicians and industrialists. In spite of having more and more pollution-control laws on the books than ever before and despite the expenditure of billions of dollars on planning, building, operating, and monitoring end-of-pipe pollution control devices, the public continued to hear of new toxic substance and hazardous waste problems. Locations like Love Canal, Times Beach or Bohpal, substances like PCBs, dioxins, or TCE, and complex conditions like acid rain, smog, or leaking underground storage tanks continue to be brought to the public's attention.

The public has become aware that exposure to these substances or conditions may cause adverse health effects ranging from eczema and respiratory diseases, to miscarriages, fetal anomalies, and to cancers of many types. Of all the maladies caused by toxic substances, the public is especially fearful of cancer, and of anything that may increase the probability of developing the dread disease in one of its various forms. Medical and environmental scientists estimate that as many as 60 to 80 percent of all human cancers are caused or exacerbated by exposure to toxic and hazardous substances. It is not surprising that people have been looking in new directions for solutions.

Within the agricultural industry, this search has resulted in the development of integrated pest management approaches that work more sensitively with the cycles of nature to prevent or minimize the build-up of pests. The result is a decreased use of pesticides for managing pests once they reach economically threatening proportions.

Within the medical profession, increasing emphasis is being placed on disease prevention and wellness promotion to help people stay healthy, rather than placing such heavy reliance on the use of drugs for controlling one's diseases once contracted.

In both the agricultural and the health-care sectors, the truth of the old adage, an ounce of prevention is worth a pound of cure, is being substantiated. Fortunately, industry's attention to pollution prevention and waste minimization are yielding similarly positive economic and environmental results.

As the case studies included in this volume illustrate, industrial leaders are developing and implementing innovative ways to reduce their production of wastes of all types and, at the same time, are increasing their profits. They are simultaneously realizing bottom-line and ecological benefits.

The 3M Corporation has been a leader in implementing what they call their Pollution Prevention Pays program, or 3P, for short. Their 3P emphasis was begun in 1975 as an alternative to the costly pollution-control-only approach mandated by governmental agencies responding to the widespread public concern about toxic and hazardous substances. The aim of 3P is to prevent the production of pollutants at their source, rather than controlling them at the end of the manufacturing process. 3M has actively involved their employees, at all levels of industrial operations, in making suggestions for ways to prevent the production of wasteful pollutants. The program has elicited impressive employee responses, saved dollars, prevented pollutants, used resources more wisely, saved energy, and advanced technology. In the first nine years of the 3P program, 1,200 employee suggestions were approved and, together, they helped the company save $192 million. At the same time, the company eliminated the discharge of more than 90,000 tons of air pollutants, 10,000 tons of water pollutants, and 140,000 tons of sludge, and prevented the production of approximately one billion gallons of wastewater. Additionally, they saved 254,000 barrels of oil annually through increased operational efficiency. (Bringer, 1984.)

Many other industrial firms are also making great strides toward total resource stewardship. The case studies included here present evidence that sound industrial management of resources results in simultaneous economic and ecological benefits.

Before proceeding with the information provided by these industrial leaders, the following sections discuss the magnitude of the problems confronting us in the area of toxic and hazardous waste management.

TYPES AND QUANTITIES OF HAZARDOUS WASTE GENERATED

Characterization and Definition of Hazardous Waste

The federal government has effectively preempted the designation and classification of hazardous wastes through the issuance of definitions pertaining to reporting requirements. The threat of financial and criminal penalties for the mismanagement of designated hazardous wastes has meant, for all practical purposes, that concerned parties have subscribed to the definition of hazardous waste as stated in the Resource Conservation and Recovery Act (RCRA).

The RCRA was designed, somewhat arbitrarily, to emphasize individual generators of hazardous waste producing in excess of 1,000 kilograms (kg) per month (2,200 pounds). More recently, with the passage of the RCRA reauthorization, this threshold reporting requirement was modified to include all generators producing in excess of 100 kg per month. Even under this definition, the estimated 3 to 8 percent of hazardous waste produced by small generators, who escape regulation, is not included in most considerations of waste management strategies.

This omission is consistent with our poor appreciation of how hazardous wastes are spawned by our consumption/disposal habits. Like the small generator, hazardous wastes disposed of by the general public are not included in the federal definition for purposes of management and waste reduction. With respect to the federal definition of hazardous waste, household hazardous waste constitutes zero percent of hazardous wastes generated.

A more comprehensive context for understanding resource use and by-product management would help us to conceptualize how the various phases of resource use are interrelated, and would be constructive in promoting resource conservation and waste reduction. Three primary phases are identified for this purpose:

1) feedstock or primary resource production, to be used as a component in further manufacturing processes (may include multiple subphases)
2) manufacture of finished consumer product
3) consumption and discard of product

To illustrate these phases take, as an example, the use of chromium. First, chromium is produced in large quantities as a feedstock for the metalplating industry. The millers and refiners of the metal characteristically generate a hazardous by-product. They have traditionally supplied hexavalent chromium to the metalplaters who utilize the element in chrome metalplating.

Some plating technologies, still currently in use, commonly

4

waste as much as 90 percent of the chromium resource to the plating rinse bath wastewater stream. (Martin, 1984.) This is the second distinct hazardous waste stream identified stemming from the use of chromium.

A third waste stream which arises from chromium use relates to the consumption and discarding of the products manufactured with chromium as a constituent. Discarded items in the public sector are commonly routed to a sanitary landfill. The majority of sanitary/municipal landfills in the United States are listed by federal agencies as potential sources of groundwater pollution. These landfills pepper the land, and are often sited adjacent to surface waters, and above drinking-water aquifers. It has been shown that years of percolating rain leach the chromium and other heavy metals, soluble organics and minerals, oils, and other toxins from the decaying buried waste and into the natural water cycle. Ordinary household wastes and other by-products of our industrialized culture contain many of the same toxic metal components, oils, solvents, caustics, and other hazardous substances found in industrial wastes.

This third waste stream will still exist despite all our mental and technical efforts to minimize wastes from primary production and manufacturing process by-product streams. This phase of hazardous waste generation is addressed only briefly in this document, as it is a psychological and sociological issue beyond the scope of this work.

There is a significant challenge in altering the consumption habits of an entire culture. The issue of household hazardous waste strikes to the root of resource use and disposal, highlighting the fallacy inherent in perceiving disposal as an actual end. The concept of disposal in a closed ecological system is, for all practical purposes, the means for a current society to avoid responsibility for total resource stewardship. Disposal shifts the burden of inefficient and wasteful resource use to future generations who will face both resource diminishment and contamination of critical life resources.

Hazardous Waste Categorized as a Function of Regulation

For purposes of regulation under the Resource Recovery and Conservation Act, wastes which exhibit characteristics of explosivity, ignitability, corrosivity, reactivity, infectivity, or toxicity are regarded as hazardous. The United States Environmental Protection Agency (EPA) lists by name some 400 specific substances which exhibit one or more of the above characteristics. However, EPA also excludes numerous substances which are known to exhibit such hazards. Although the regulatory and legal framework has introduced some loopholes, incomplete toxicological data on many chemicals contributes to uncertainty in the designation of hazardous wastes. The EPA maintains a procedure whereby wastes can be listed or delisted as hazardous based on scientific findings.

There are a variety of classification schemes frequently used in characterizing the hazardous waste stream. Most often, the wastes are characterized by their chemical properties in greater or lesser detail. Other useful categorizations include wastes managed on- or off-site, potential hazard, waste generated by industry group, or the physical properties of the waste, such as whether it is solid, liquid or gas, and its moisture content. Such properties are important to consider both in seeking broad policy goals for pollution prevention, and in designing appropriate waste minimization and detoxification strategies.

Hazardous wastes are most popularly classified as encompassing variations of the following categories:

o metallic inorganic
o non-metallic inorganic
o halogenated organic
o non-halogenated organic
o cyanides
o pesticides
o corrosives
o explosives

The most reliable information regarding the amount of waste generated in each of the hazardous waste classifications is available through the federal and state hazardous waste management programs, most notably, the EPA. The information is not exact because of limitations inherent in the regulations.

In their May 1985 report on hazardous waste management, the Congressional Budget Office (CBO) estimated that about 266 million metric tons (MMT) of hazardous wastes were generated by industries operating in the United States in 1983. This estimate is within a 95 percent statistical confidence interval, and is in keeping with other estimations of hazardous waste, generated in 1981, made by the EPA and the Office of Technology Assessment (OTA). The OTA and EPA estimates ranged from a low of 132 MMT to a high of 395 MMT, with the most probable amount being 264 MMT. The CBO estimates waste generation growth of 1 percent per year through 1990 if no significant incentives are implemented to encourage waste reduction.

Of the 450 four-digit classifications of industry in the Standard Industrial Classification System (SIC), CBO reports that 70 industries generate about 95 percent of the hazardous wastes reported in the United States. These 70 industries are categorized according to the twelve groupings listed in Table 1. Data on the amounts of waste generated, and the amounts managed on- and off-site are presented in Table 1 for these major industries.

A significant distinction between on- and off-site hazardous waste management is largely visible in the concerns of community and other advocacy representatives who question the risks of transportation of hazardous wastes, and of their concentration at centralized management facilities. This is especially true for disposal facilities. Deep-well injection, ocean dumping, landfilling, and related land-based treatments all carry their characteristic hazards to local environments. Although these same practices are performed on-site, on-site facilities have historically received less attention from concerned citizens. In contrast, nearly every state has seen a proposed hazardous waste management facility fought, and often blocked, by community and public health advocates.

Another distinction between on- and off-site facilities, of perhaps greater import for waste management, is that a waste stream which can be properly managed on-site brings the loop of total resource management under the increased control of the resource consumers. Waste by-products which are managed on-site are generally more amenable to detoxification, recycling, or reuse.

Factors which may influence the ability of an industry to treat its wastes on-site are highly variable with respect to specific industry dynamics.

They are:

Waste Factors

1)   ease of recycling or reuse
2)   ease of detoxifying the by-product
3)   amount of waste generated

and

Business Factors

1)   research and development potential of the business
2)   capital improvement funds available
3)   cost of treatment technology
4)   psychological preparation
5)   availability of off-site waste management
6)   potential joint and several strict liability claims

7

TABLE 1

WASTE GENERATED AND MANAGED ON- OR OFF-SITE IN 1983
(estimated quantity in thousands of metric tons)

MAJOR INDUSTRY	ESTIMATED QUANTITY	% OF TOTAL	% ON-SITE	% OFF-SITE
Chemicals & Allied Products	127,254	47.9	99	1
Primary Metals	47,704	18.0	98	2
Petroleum & Coal Products	31,358	11.8	99	1
Fabricated Metal Products	25,364	9.6	98	2
Rubber & Plastic Products	14,600	5.5	80	20
Miscellaneous Manufacturing	5,614	2.1	83	17
Nonelectrical Machinery	4,859	1.8	88	12
Transportation Equipment	2,977	1.1	83	17
Motor Freight Transportation	2,160	0.8	11	89
Electrical/Electronic Machinery	1,929	0.7	90	10
Wood Preserving	1,739	0.7	100	0
Drum Reconditioners	45	less than 0.1	11	89

SOURCE: Hazardous Waste Management: Recent Changes and Policy Alternatives. Congressional Budget Office, May 1985.

An alternative way of viewing the waste generation figures is to organize them according to the quantity of each waste type generated. Table 2 lists a mean quantity for each waste category taken from the upper and lower values of the estimated range of waste generated for that category.

The case studies of low- and non-waste technologies included in this book are listed by Standard Industrial Classification (SIC) code, with abstracts cross-indexed both by pollution prevention technology and waste type. Waste type is usually more suitable for choosing the technology appropriate for a given waste stream, since industry wastes vary qualitatively, and are often generated by a variety of industrial processes.

The Standard Industrial Classification code is a categorizing system used to classify industries by related fields of function. The complete index, entitled Standard Industrial Classification Manual 1972, is available in most libraries or from the Superintendent of Documents, U.S. Printing Office, Washington, D.C.

THE COSTS OF HAZARDOUS WASTE GENERATION TO INDUSTRY, SOCIETY, AND THE ENVIRONMENT

The costs of management and mismanagement of hazardous waste have been high for society and the environment, as well as for industry. The national bill for hazardous waste management and cleanup of waste disposal sites is estimated by federal agencies to exceed $6 billion a year. Under the expanded RCRA regulations adopted in 1985, future costs are likely to be higher.

The cost of waste management is particularly odious to industry, which enjoyed an element of freedom in externalizing this cost until the late 1960s and early 1970s, when federal pollution control laws were enacted. Since that time, waste management costs have increased dramatically. The 70 largest waste generating industries, identified by the CBO, incurred total costs of an estimated $5 billion (the median estimate) to treat and dispose of their hazardous wastes in 1983.

The 1985 RCRA amendments are expected to add an estimated 46 percent to industry's compliance costs by the year 1990. Such an increase would raise the costs of compliance for the entire industry from an estimated $5.8 billion in 1983, to as much as $11.2 billion in 1990, depending on the degree of success in waste reduction. As these costs swell to even larger percentages of product value added, corporations will undoubtedly seek relief, as production rates will not experience an attendant rise

TABLE 2

ESTIMATED NATIONAL GENERATION OF HAZARDOUS WASTE IN 1983
RANKED BY WASTE CATEGORY
(estimated quantity in thousands of metric tons)

WASTE TYPE	ESTIMATED QUANTITY	% OF TOTAL
Nonmetallic Inorganic Liquid	82,261	31
Nonmetallic Inorganic Sludge	28,061	11
Nonmetallic Inorganic Dust	21,120	8
Metal-containing Liquid	19,760	7
Miscellanous Waste	15,415	6
Metal-containing Sludge	14,497	6
Waste Oil	14,249	5
Halogenated Organic Solvent	12,130	5
Metallic Dusts & Shaving	9,784	4
Cyanide & Metal Liquid	7,383	3
Contaminated Clay, Soil & Sand	5,461	2
Nonhalogenated Organic Solid	4,578	2
Dye & Paint Sludge	4,236	2
Resins, Latex & Monomer	4,018	2
Oily Sludge	3,734	1
Halogenated Solvent	3,479	1
Other Organic Liquid	3,435	1
Nonhalogenated Organic Sludge	2,242	1
Explosives	720	*
Halogenated Organic Sludge	715	*
Cyanide & Metal Sludge	557	*
Pesticides & Herbicides	26	*
Polychlorinated Biphenyls	1	*

* - less than 1 percent

SOURCE: Hazardous Waste Management: Recent Changes and Policy
         Alternatives. Congressional Budget Office, May 1985.

from the waste management investment dollar. Many corporations have met this challenge through the modification of processes to reduce and eliminate waste, rather than investments in pollution control.

The cost-weighted average for all industrial waste management was aproximately 1.3 percent of the value added. As is evident from the figures in Table 3, compliance costs for some industries appear high enough to erode a significant part of their profit margin. As costs rise, some plants will be forced to close if costs cannot be passed along to customers. Higher prices, in turn, could reduce demand for domestically produced products, since overseas production facilities may not operate under such strict regulations. However, as our case studies document, industrial emphasis on pollution prevention is presenting economically viable alternatives to these increasing hazardous waste disposal costs.

In addition to the costs incurred by industry directly, other segments of society have also experienced high costs associated with hazardous waste. Although some costs are quantifiable, often they are of a subjective nature or are too interwoven with other factors to be quantified. There is no reliable national estimate of costs to private individuals which include medical expenses due to acute or chronic exposure to hazardous wastes. This is largely due to the lack of a data base from which to compile the information, and to the fact that pollution victims are not always aware of their disease etiology. In the case of the chronic toxicity of Love Canal, the full extent of a victim's injuries may not be known for 10 to 20 years or more, because chronic diseases, such as cancer, heart disease, neurological disorders, and reproductive problems often have long latency periods. To compound the difficulty of assessing health costs there are confounding variables which undermine certainty in cause and effect studies.

The OTA report, Protecting the Nation's Groundwater, documents increasing evidence of land and water contamination from toxic substance burial sites, pesticide application, and chemical and gasoline storage tanks. Such releases of toxics into the environment can only lead to increased exposure for individuals. Indications are that costs in this sector are growing. For example, Love Canal victims were recently awarded a $20 million settlement.

Burial and other land-based hazardous waste management approaches do not factor in the long-range costs to society commonly associated with them. The OTA has estimated that the cost of cleaning up a land-based facility which is contaminating groundwater is typically 10 to 100 times the cost incurred had

TABLE 3

EXPENDITURES FOR HAZARDOUS WASTE MANAGEMENT & PERCENT OF VALUE ADDED FOR MAJOR INDUSTRIAL GROUPS
(in millions of 1983 dollars)

INDUSTRY GROUP	EXPENDITURES	VALUE ADDED	% VALUE ADDED
Wood Preserving	56	432	13.0
Fabricated Metal Products	899	19,282	4.7
Chemicals & Allied Products	1,544	48,058	3.2
Rubber & Plastic Products	798	28,270	2.8
Primary Metals	1,243	44,770	2.8
Miscellaneous Manufacturing	267	15,384	1.7
Nonelectrical Machinery	254	27,563	0.5
Electrical/Electronic Products	156	46,507	0.5
Petroleum & Coal Products	136	29,327	0.5
Transportation Equipment	191	68,904	0.3
Motor Frieght Transportation	299	No data available	
Drum Reconditioners	6	No data available	

SOURCE: Hazardous Waste Management: Recent Changes and Policy Alternatives. Congressional Budget Office, May 1985.

the wastes been treated in a more responsible manner in the first place. EPA estimates that the cost to clean up and contain polluted groundwater is $5 to $10 million per site. The cost to totally reclaim a polluted aquifer may be ten times that.

While the EPA has placed 2,000 sites on their National Priorities List, OTA estimates that 10,000 or more sites may require clean-up by Superfund. (Superfund Strategy, Office of Technology Assessment, March 1985.) Thus, a cleanup costing as much as $100 billion already looms before us.

Even the limited discharge of dilute toxic waste streams meeting federal effluent standards exacts a cost to society, in addition to the potential adverse public health effects. When these wastes are discharged to a municipal or other privately operated wastewater/sewage treatment plant, the toxic constituents can disrupt the organic treatment systems by killing the treatment organisms. Equally important is the concentration of the industrial waste in the sludge formed in treatment. Case studies included in this volume indicate that sewage sludge can be easily processed to form a valuable soil amendment high in nutrients, and of good texture. This recyclable product is grossly under-used because of its large scale contamination with toxic industrial by-products. Thus, this particular cost to society is retarding an entire for-profit and socially beneficial industry.

In addition to the expenditures relating to hazardous wastes noted above, there are the costs incurred by various segments of society battling one another for what are claimed to be "best interests". Additional billions of dollars are spent yearly by government to regulate industry, by industry to lobby the public and government, and by environmental and other special interest groups to promote their respective "ecologically sensitive" agenda. An effort to pool these resources toward mutually agreeable goals of pollution prevention and waste minimization could be much more productive. A society working together to solve common problems, as opposed to the current in-fighting and squandering of resources, will allow us to proceed with the task of practicing total resource stewardship.

POLLUTION PREVENTION-WASTE MINIMIZATION: ESSENTIAL FOR ECONOMIC AND HUMAN HEALTH

The best hope for moving our industrialized society beyond the continual creation of additional superfund sites and rapid escalation of waste disposal costs, is to implement pollution prevention-waste minimization approaches throughout all the industrial sectors of our society. Net profitability can thereby be increased at the same time that new, stricter waste management regulations are implemented.

The case studies included in this book illustrate that pollution prevention and waste minimization are already technically feasible and have been implemented in many industrial sectors. The reductions in wastes produced range from 20 to 98 percent. Additionally, some firms have reduced the degree of hazard of their wastes so that they no longer produce or release any hazardous wastes.

Payback periods for these waste minimization investments range from immediate to five years. Concomitantly, since each firm is producing and releasing fewer hazardous pollutants, they have reduced the hazards to their workers and to the environment. As a result, these firms will experience fewer long-term waste liability and victim compensation claims.

In assembling the information for the case studies, several industrial leaders commented that the motivation to investigate waste minimization alternatives was three-fold:

1) environmental regulations mandating stricter effluent guidelines

2) rapidly increasing costs of previously inexpensive "waste disposal" approaches such as landfilling

3) enactment, or potential enactment, and enforcement of regulations pertaining to joint and several liability, worker and citizen right-to-know laws, and victim's compensation

These three clusters of regulations and situations were the stick which initially received the attention of the industrial leaders, and prompted them to look for waste minimization alternatives. Once the carrot of financial and environmental benefits of utilizing these approaches became apparent, one industrial leader was led to express his sentiments as follows:

> I wish we had been forced to look at waste reduction in this comprehensive way ten years ago. We are now saving $1 millon per year. Just think what we could have done if we had started earlier!

Other industrial leaders made similar statements. Some leaders saw a close relationship between the role of increasing energy costs that forced society to be more energy efficient, and the rising health and environmental costs forcing industry to be more efficient in the management of its total resources, including its toxic and hazardous substances.

Pollution prevention and waste minimization approaches can help our society avoid the pitfalls referred to in the following quote:

> To waste and destroy our natural resources instead of increasing their usefulness, will undermine the very prosperity which we are obligated to hand down to our children, amplified and developed.
>
> -Theodore Roosevelt, 1907

CHAPTER 2

Pollution Prevention—Waste Minimization Concepts and Approaches

OVERVIEW OF POLLUTION PREVENTION AND WASTE MINIMIZATION

The goals of clean air, fishable and drinkable water, and safe workplaces have been evolving throughout the last 15 to 20 years. As each new environmental crisis emerged, lawmakers responded with new regulations to "control" the problem. But as stated before, the consequences of the piecemeal, medium-by-medium pollution control efforts have been less than satisfactory. Consequently, some industrial leaders, in looking for better ways to comply with the environmental regulations, found they could more efficiently achieve the regulatory goals by placing primary emphasis on methods that prevent the production of pollutants, rather than attempting to control them once produced.

In the fall of 1984, the U.S. Congress passed and President Reagan signed the RCRA reauthorization. Several amendments are designed to drastically reduce the quantities and types of wastes permitted to be buried in landfills, and to encourage industries, small and large, to undertake efforts to minimize the production of hazardous wastes. Furthermore, each generator must certify that a waste minimization program is in operation in its plant. Each generator is also required to quantify the reduction in waste volume and toxicity achieved each year. Each must also certify that the method of treatment, storage, or disposal is a practicable method which minimizes the present and future threat to human health and the environment (Resource Conservation and Recovery Act of 1976, as amended, published in the Congressional Record, October 3, 1984 & October 11, 1984.)

Thus Congress, in reauthorizing the RCRA, mandated that industry employ waste minimization and toxicity reduction approaches and decrease its reliance upon the environmentally dangerous methods of land interment.

The authors of this book submit, for your use, case studies of the economic successes achieved by approximately 50 industries through reducing the quantity of their hazardous and non-hazardous wastes, and decreasing the toxicity of their hazardous wastes.

In the case studies, each organization's personnel usually performed environmental or waste audits to determine the types and quantities of wastes being produced. They then utilized approaches, such as process modifications, and component substitutions, to prevent or substantially reduce the quantity of wastes generated.

Next, by segregating the waste streams and analyzing them, the companies found many ways to reuse or recycle the constituents on- or off-site.

Following this phase, the remaining wastes were detoxified by chemical, biological, or physical means. The residues from these processes were ideally safe for release into the environment. If, however, there are any remaining hazardous components, it should be mandated they be placed in retrievable storage. Industry and society should be charged with underwriting research to develop methods for the further detoxification of these remaining materials. Once developed, such methods should be employed to detoxify the materials held in storage.

As a guide to the reader, Table 4 presents the hierarchy of pollution prevention-waste minimization strategies.

Many of the industries that have been most successful in following the different steps in this hierarchy have utilized an incentive awards program to encourage their employees to offer ideas for solving the hazardous waste management problems of the company. Such programs might also become incorporated as part of other companies' waste minimization efforts.

REVIEW OF INDIVIDUAL COMPONENTS AND PROCESSES APPLICABLE FOR POLLUTION PREVENTION AND WASTE MINIMIZATION

In the following sections of this chapter, individual components and processes of the pollution prevention-waste minimization hierarchy are addressed. These provide an introduction to the approaches the industries have found useful.

Environmental Audit/Waste Audit

The waste audit is an offshoot of the environmental audit. Because there are no strictly observed protocols for performing an audit, it could be argued that the two are one and the same; both examine the processes a facility regulates in order to maintain environmental quality and to manage the dispositon of industrial by-products. Environmental audits are, in the most general terms, used to determine the adequacy of an organization's current practices in complying with environmental regulations.

TABLE 4

HIERARCHY OF POLLUTION PREVENTION-WASTE MINIMIZATION
STRATEGIES USED BY INDUSTRY TO EFFICIENTLY MEET ENVIRONMENTAL
QUALITY STANDARDS

===

A. Environmental Audits & Waste Audits

B. Pollution Prevention

C. Waste Minimization

D. Resource Recovery, Reuse & Recycle

    1. on-site
    2. off-site

E. Waste Detoxification & Destruction

    1. neutralization
    2. chemical
    3. biological
    4. thermal

F. Retrievable Storage

    1. short-term
    2. long-term encapsulation &
       solidification

===

A well-designed and properly performed audit can provide an organization's leadership with important information. Audits can indicate problems in processing which appear chronic or lead to remedial work and/or fines for non-compliance with environmental regulations. In this regard, they can identify repetitive compliance problems, and ensure emergency preparedness. They are useful in assisting plant management in certifying compliance with regulations. Information obtained through an audit may be important in preparing the disclosure statements required by the Securities and Exchange Commission. Finally, they enhance the environmental awareness of management-level personnel, and provide an opportunity for positive public relations on environmental concerns. Audit design may have a variety of foci. Costing out compliance strategies, determining existing or contingent liabilities, and providing a planning mechanism to determine future costs associated with compliance may be independently or jointly designed into an audit.

The waste audit, however, can be designed as a detailed examination of all aspects of process design and material flow which relate specifically to waste management. A subsequent plan may be developed for the management of waste, thus optimizing its environmental sensitivity and economic benefits. The waste audit can be used to identify appropriate technology and innovative methods which minimize the organization's costs and waste generation. It has been used to successfully identify the raw materials likely to result in hazardous waste generation before they are purchased, thus allowing for the use of alternative substances. The inventory of wastes generated can be useful in planning waste segregation for the recycling of certain waste streams back into processes or to external markets. In one corporation a research engineer is assigned to review audit information to seek opportunities for reducing pollution and costs through process or material changes. Audit information is also useful for technology transfer and employee waste-reduction incentive programs. Use of the audit information can help stimulate personnel to think in terms of proactive pollution prevention and waste reduction, rather than focusing solely on pollution control, as has been the case in the past. Properly performed and utilized, waste audits will assist industry in setting goals for waste minimization and hazard reduction, and in quantifying their progress in achieving those goals.

On-Site Waste Reduction and Pollution Prevention

Following the waste audit, instituting strategies for the reduction of wastes on-site is the initial step in a good waste management system. Waste reduction is the first phase of the waste management hierarchy. On-site prevention of the production of wastes can often be easily and inexpensively accomplished. Many opportunities may exist in a facility which has never had a waste audit performed. Improved housekeeping techniques, such as routine checks for leaking or inadequately maintained equipment, optimizing flows of materials (process contols) and checking the mass balance of flows in and out of the processes to detect unusual losses, can significantly contribute in identifying opportunities for waste reduction.

Other, more involved initiatives include substitution of low- or non-waste generating raw materials for those which generate problematic waste by-products. An example of this approach, gaining increased use, is the substitution of water-based solvents for oil-based solvents which require special care in their handling as by-products.

With the development of energy- and resource-efficient technologies, there are opportunities for the modification of process equipment to promote waste minimization. Either the addition of closed-loop cycling of materials, or the introduction of more efficient processing technology may result in decreased waste generation, and lower production costs.

Other opportunities for on-site waste minimization, though also applicable at off-site facilities, are the reclamation of components in the waste, and recycling. These may be realized as process modifications, but often require additional processing beyond that specific to the manufacturing process. Variations of this include the separation of waste streams to make recovery of recyclable materials easier. For example, processes may be applied which separate components of wastes to allow partial or complete reuse of the materials, and the interindustry exchange of by-products, from the generator to a manufacturer, may utilize this material as an inexpensive feedstock replacement or supplement. Waste treatment facilities are finding that the recycling of waste streams and their components generates revenue, and reduces the amount of residue requiring additional management.

Solvent recycling, performed on-site at many manufacturing facilities, has been found to be particularly advantageous. The advantages cited for this are:

o elimination or reduction of hazardous waste disposal costs
o elimination or reduction of potential long-term liability from disposed (buried) hazardous wastes
o rapid return on capital equipment investment
o no federal EPA permit required (see Federal Regulation 40, Part 261)

o reduction of storage of hazardous material on-site,
  recovery is completed as material is being generated
o assurance that solvent reclaimed in-house is not
  contaminated with other outside solvents
o tremendous savings in production costs by reusing
  solvents

On- and Off-Site Resource Recovery, Reuse, and Recycle

After the opportunities for waste minimization and recycling
have been realized, on-site and off-site treatment of the
remaining substances, to detoxify or destroy them, is desirable.
Detoxification processes are biological, chemical, physical, or
thermal.

The appropriate use of biological treatment is limited to
those wastes which can be biologically degraded. The technology
for biological treatment has been in service for years, and
includes most elements of the standard sewage treatment plant,
trickling filters, aerated lagoons, and activated sludge, as well
as rotating biological contactors, deep-shaft oxidation,
landfarming, and the application of biotechnologically produced
strains of microorganisms.

Most of these techniques require the periodic removal of
sludges. Ideally, the sludges will be applied to land as an
amendment. Because the degraded waste residues will eventually
be worked into the land, it is important that wastes treated in
this manner are fully biodegradable, and are not contaminated
with heavy metals or other toxics with long residence times in
the natural environment. Industrial by-products commonly treated
in this fashion are oils, phenols, alcohols and sulfides, as well
as vegetative by-products, such as are associated with breweries
and pulp mills.

Chemical and physical treatments involve a wide variety of
technologies ranging from simple separation procedures, such as
settling tanks and filtration, to precipitation and delicate
chemical reactions. Dilute wastes are usually treated so as to
concentrate them on-site prior to shipment off-site for further
treatment. In many cases, concentration processes can be applied
as recovery techniques for materials and water. Ultrafiltration,
reverse osmosis, electrolytic recovery, distillation,
precipitation, flotation, centrifugation, ion exchange, and a
wide variety of other filtration techniques are illustrative of
separation and concentration methods.

Waste Detoxification and Destruction

The neutralization or destruction of toxic by-products by chemical reactions is a research chemist's garden of possibilities. The neutralization of acidic by-products with alkaline wastes is commonly practiced. Chlorine stripping, cyanide destruction, and a wide range of precipitation applications are other examples. The limitations to the application of chemistry for waste treatment lie within the inclination and creativity of process engineers and chemists.

Thermal destruction has been identified by many federal agencies as an important technology for waste treatment likely to see expanded use. Waste destruction has been found to be very satisfactory, with the added incentive of often providing for the recovery of the waste's energy potential through the cogeneration of electricity and steam.

Thermal destruction is classified as either incineration or pyrolysis. Incineration is the combustion of waste by-products in the presence of excess oxygen. With pyrolysis, the materials to be treated are combusted in an oxygen-deficient environment. Organics are decomposed into simpler molecules, often suitable for recovery, and ash. In both of these methods, stack emissions will ideally be carbon dioxide and water, with bottom ash as another residual. Sometimes, however, the residual contains toxic components which require further management. For these reasons, the use of long-term storage and stabilization strategies is recognized as essential to a complete waste management system.

Long- and Short-Term Retrievable Storage

It must be emphasized that, as the least desirable element of a total waste management system, long-term storage of waste treatment residuals is appropriate only for those materials which, with presently available technology, cannot be further reduced in volume or toxicity. This is an important contrast to the current use of land disposal practices as an inexpensive preliminary or sole management strategy. Hazardous treatment residuals in long-term storage should ideally be made retrievable, so that, when technology is available for the detoxification or recovery of the waste constituents, they may be so treated. Such provisions are required in order to exclude from the concept of long-term storage the erroneous notion of "disposal".

Technologies are being developed to ensure the maximum security of stored wastes. Solidification techniques rely on the fixation and encapsulation of waste constituents to physically encase or immobilize their toxic components. Current techniques include fixation strategies which mix the wastes into cement, pozzolanic material, or polymers, and other stabilization techniques which seal the waste into a matrix. Encapsulation is also performed, largely utilizing petroleum-based resins. In this process, wastes are containerized in a seamless shell to minimize and delay contact of the waste with the environment.

While encapsulation or retrievable storage may be the necessary management option for some wastes in the near term, in the longer term, Congress and environmental necessity dictate an increased reliance on pollution prevention and waste minimization, with a decreased reliance on landfilling.

The new RCRA "hammer" dates for the introduction of strict landfilling provisions will strongly discourage the use of landfills and other land interment and storage facilities. Currently, surface impoundments are the second most prevalently used form of land containment strategy after landfilling. They received approximately 50 MMT of hazardous waste in 1983. The EPA estimated that 70 percent are unlined, and that 90 percent pose a threat to the local groundwater.

There are other indications of the unsuitability of land burial and containment practices. A recent survey performed by the House Committee on Energy and Commerce staff found some indication of groundwater contamination at 559 of 1,246 RCRA-authorized facilities. The preamble to EPA's land disposal regulations published in the Federal Register, Volume 47, Number 143, reads, "most land disposal units, however well designed, will eventually leak after closure to some extent."

Clearly, Congress has been given adequate impetus to act to protect the public health from harm caused by inappropriate waste management strategies. Congress is continuing to seek ways of promoting waste reduction as the premiere method for managing hazardous wastes. There are indications that "waste-end" taxes are being considered as one approach to discourage waste generation. The CBO suggests that "waste-end" taxes would significantly encourage hazardous waste reduction, and support shifts to new technologies desirable under the RCRA amendments.

Although waste reduction is the most desirable and advantageous waste mangement strategy, it is but a part of the entire range of options which combine to optimize a cost-effective mangement system. A simple and idealized system progresses from initial efforts to minimize waste generation, to recycling and recovery of materials, to application of treatment processes for detoxification and destruction of the waste, to long-term storage and immobilization of residuals.

Wastes to Resources

CHARACTERIZATION OF WASTES NOT PRODUCED, RECOVERED, OR RECYCLED AS A CONSEQUENCE OF STRATEGIES DOCUMENTED IN THE CASE STUDIES

Some wastes are more readily not produced, or are more easily recovered, than others. However, very little data that quantify the degree of reduction that has been achieved are available. Information is oriented toward wastes generated, as opposed to wastes not produced.

Halogenated and Non-Halogenated Solvents

Our studies show that organic solvents are being replaced, in some industries, by water-based solvents. In other instances, the organic solvents are being extensively recycled and reutilized. Some facilities recycle 60 to 90 percent of their solvents. Typical payback periods for the capital investment in recovery facilities and equipment are usually less than a year.

In some facilities, the very heavily contaminated solvents and still bottoms are blended with other fuels, and used in boilers as energy sources.

The combination of technically feasible and economically beneficial solvent management approaches should result in 80 percent recovery of solvents throughout industry within the near future. This represents a substantial savings to the solvent user and a dramatic reduction in potential environmental risk.

Waste Oils, Machining Oils

Rerefining of used oils is becoming increasingly feasible. Some firms have set up facilities to reprocess their used motor oils, machine tooling oils, and other industrial process oils. Those companies generally report payback periods on their oil reprocessing investments of two to five years.

Some firms are blending their used oil with fuel oil and incinerating it for its energy value.

From our data, it appears that much progress can be made through expanded rerefining and reuse of oils at small in-plant facilities, or at larger regional reprocessing plants. Additionally, waste oils, oily sludges, and related high BTU-value organic wastes, if incinerated in a proper facility, can serve as valuable energy sources. Such practices will utilize our energy resources more efficiently, and minimize environmental risks from the wastes.

Metals and Metal-Containing Sludges

Depending upon the element, its concentration, toxicity and value, an array of reuse and recovery approaches are technically and economically feasible.

Several companies, illustrated in our case studies, have installed electrolytic processing units for the recovery of silver, copper, and other metals. Others have incorporated ion exchange columns to prevent the wasteful release of copper, nickle, and chromium. Payback times on these process modifications frequently range from 6 to 18 months.

An array of processes, designed to reduce water usage, or to close the loop in process water circulation, are also making it economically feasible to minimize the production of metal wastes. Savings in water costs, metal costs, energy costs, and waste management costs result in payback periods ranging from 5 to 20 months.

Several firms have made process modifications to ensure that their metal-containing sludges are readily reusable. One North Carolina firm, for example, modified its process to decrease contamination, and now finds it economically profitable to ship its 40-percent copper sludge to Denmark for copper recovery. Payback on these types of capital investments range from 3 to about 14 months.

A tremendous potential exists for waste minimization and resource recovery in the fields of metal use. New technology to promote the recovery of metal from sludge wastes is in the late stages of development. Many electroplaters and metalworkers have indicated that they are storing their sludges in the hope that their metal constituents will soon be economically recoverable. We anticipate that the RCRA reauthorization amendments which encourage waste minimization and discourage landfilling, along with the technically sound and economically successful experiences of others, will stimulate the development of such technology, as well as encourage other metal-using firms to implement these strategies.

Pesticides

By careful segregation of pesticide waste streams, one pesticide manufacturer in our study eliminated all waste produced in the process and increased its efficiency by returning the previously wasted substances to the appropriate process streams. Waste stream segregation is an important first step in developing reuse approaches in many industrial processes.

WASTE REDUCTION PROJECTIONS

The foregoing statements, based on information from our industrial associates, suggest that much waste reduction and resource recovery is technically and economically feasible at this time. While many firms are already utilizing waste minimization approaches, many other firms have not incorporated these concepts, procedures, and technologies into their operations.

The authors hope the following case studies will foster many industrial process modifications, resulting in waste reduction and economic benefits for all. While progress can be made by firms in practically every industrial sector, there are waste streams that presently provide serious obstacles for minimization. These particularly difficult waste streams should be identified and researched for ways to prevent or reduce them. Additional research devoted to detoxification approaches is also urgently needed.

Several university-industry research centers have been, or are being established in the United States. They will be well-suited to devise new waste elimination and detoxification technologies. These centers will certainly be in the vanguard, moving our society toward ecologically and economically sound management of our resources.

CHAPTER 4

Case Studies Organized by Standard Industrial Code

Case Study 1

TYPE OF INDUSTRY: Swine Production

NAME OF FIRM: The Choplin Farm (919/556-2383)

LOCATION: Wake Forest, North Carolina

SIC: 0213 & 0115

```
*********************************************************************
*                                                                   *
*                        MODIFICATIONS                              *
*                                                                   *
*    1)  THE USE OF SWINE WASTES AS A FERTILIZER                    *
*                                                                   *
*    2)  APPLICATION OF A COMPREHENSIVE SOIL AND WATER              *
*        CONSERVATION PROGRAM                                       *
*                                                                   *
*********************************************************************
```

Abstract

In 1982, the Choplin Farm saved $8,000 in fertilizer costs by
spraying swine wastes on crops, using a standard irrigation gun.
They also reduced organic material losses by 1,670 percent, total
nitrogen losses by 490 percent, and total phosphorus losses by 525
percent through the integrated use of conservation tillage,
parallel terracing, grassed waterways, and cover crops. The reuse
of these agricultural wastes, along with the management initiative
to incorporate conservation practices in the farm operation,
resulted in corn crop yields that were 300 percent above the
county average.

ANNUAL COST SAVINGS: $107,000

PAYBACK PERIOD: 6.5 years

TIME TO IMPLEMENT: Information not provided.

TRANSFERABILITY: Applicable to animal-farming operations
                         in conjunction with land-farming.

ENVIRONMENTAL AND
HEALTH BENEFITS: Conservation of nutrients and water;
                         decrease in soil erosion; reduction in
                         nonpoint sources of pollution; decrease
                         in risk of drinking-water
                         contamination.

Project Description

The Choplin Farm is a swine-raising operation. Throughout the year, animal wastes are collected in a slurry pit; in early spring, soil samples are taken for analysis. Based on the results, particularly the nitrogen and phosphorus concentrations in the slurry, the wastes are diluted, with water from ponds used to store runoff, to the desired concentration. The diluted swine-waste is pumped underground, to stations distributed around the farm, and is applied periodically to the crops by a traveling hard-nose sprayer. During 1982, this practice saved $8,000 in fertilizer costs, and provided environmentally sound management of the wastes.

In concert with swine-waste application, the following best management practices are being used to decrease nonpoint source pollution:

1) Soil and water conservation practices, such as grassed waterways, parallel terraces, field borders, winter cover crops, and conservation tillage, are employed to reduce sediment runoff.

2) Three ponds are available to store stormwater runoff for seasonal irrigation needs.

3) Soil is tested annually to assess nutrient and lime requirements.

4) Irrigation is scheduled based on crop and soil moisture conditions.

As a result of these practices, the farm operation reports that:

1) Stormwater runoff has been reduced from 176,00 gallons per acre to 93,000 gallons per acre.

2) Sediment losses have been reduced from 14.7 tons per acre to 0.05 tons per acre.

3) Organic material lost from soil has been reduced from 1,370 pounds per acre to 82 pounds per acre.

4) Total nitrogen lost from soil has been reduced from 38.2 pounds per acre to 7.8 pounds per acre.

5) Total phosphorus lost from soil has been reduced from 12.6 to 2.4 pounds per acre.

These modifications have improved crop yields. The Choplin Farm produced North Carolina's fourth-largest corn yield in 1981 -- 214 bushels per acre.

PROCESS DIMENSIONS

==

WASTES GENERATED	Reduced sediment, organic, and nutrient loading to surfacewater
PRODUCTIVITY	Improved
PRODUCT QUALITY	Improved
DOWN TIME	No change
FACILITIES	No change
EQUIPMENT	$71,000 one-time cost
RAW MATERIALS	$8,000 saved per year
WATER	83,400 gallon saved per acre
ENERGY	Increased savings
WASTE DISPOSAL	Decrease in disposal problems
POLLUTION CONTROL	Decrease in control costs
PERSONNEL/MAINTENANCE	Detailed information not provided
NET BENEFITS	Increased yield, reduced costs, decreased soil erosion, and surface-water pollution

==

For more information, refer to:

Agricultural Nonpoint Source Control Case Studies In North Carolina, IX, Wake County Demonstration Farm, by J. B. Atkins, Biological and Agricultural Engineering Department, North Carolina State University, Raleigh, North Carolina, May 1984, 119 pp.

"Irrigation/Hog Waste Slices Fertilizer Bill 80 Percent," by F. Jeter, in Progressive Farmer, Febuary 1983, p. G-14.

"No-Till Makes the Grade," by D. L. Comis, in Soil and Water Conservation News, July 1983, p. 6-7.

"Pollution Control and Production Efficiency," by J. B. Atkins, in Extension Review, Spring 1983, p. 47.

Case Study 2

TYPE OF INDUSTRY: Molybdenum Mining

NAME OF FIRM: Climax Molybdenum Company (Lauren Ames
303/279-7636)

LOCATION: Leadville, Colorado

SIC: 1061

MODIFICATION

**WATER TREATMENT PROCESS WITH MOLYBDENUM RECOVERY BY
ION EXCHANGE**

Abstract

Climax Molybdenum Company modified the natural drainage system surrounding their tailings impoundment to reduce the quantity of water requiring treatment, and to recover soluble molybdenum from the tailings runoff. Interceptor canals were built upstream to divert natural runoff waters around the tailings and into natural streams. The storm water runoff from the tailings is now undiluted by upstream runoff, and more amenable to recovery of disolved suspended material. It is passed through a countercurrent ion-exchange unit in order to remove the molybdenum, which the company then recovers and sells. Other heavy metals are separated from process water through the use of the Swift Lectro-Clear process. This process removes over 90 percent of the iron, manganese, zinc, and copper from the tailings stream by a coagulative, electrolytic process. Cyanide, used as a copper depressant in the regrind plant, is 96-percent removed.

ANNUAL COST SAVINGS: Information not provided.

PAYBACK PERIOD: Information not provided.

TIME TO IMPLEMENT: Information not provided.

TRANSFERABILITY: Applicable to mining operations and
other industries using water, which

becomes contaminated with metal and
cyanide, as their process solvent.

ENVIRONMENTAL AND
HEALTH BENEFITS: Reduced threat of ground- and
                                 surfacewater contamination from heavy
                                 metals; conservation of water and
                                 molybdenum; decreased health risks due
                                 to heavy metal and cyanide pollution.

Project Description

Since the 1930s, the Climax Molybdenum Company has operated a
water recycling system; active tailing dams and water storage
facilities are used to collect millwater, and that water is
returned for reuse in the plant. The system provided total
containment and capture of wastewater, except during the annual
spring snowmelt. In order to keep the system in balance, tailing
water was bled into the nearby creek at a controlled and monitored
rate; in this way, the quality of the stream was maintained.

In order for the mine to discharge water continually,
the water had to be treated to remove molybdenum, copper, zinc,
iron, manganese, and cyanide. To implement the treatment system
economically, Climax Molybdenum Company, aided by Amax Extractive
Research and Development, studied the drainage system surrounding
the tailing impoundment area. From the data gathered, they were
able to reduce the amount of water that required treatment by
building a series of interceptor canals to bypass runoff waters
around the tailings system and into the natural streams. In
building the canals, the company also determined how much water
had to be treated in order to maintain the system's integrity.

The water to be treated is initially passed through a
standard countercurrent ion-exchange system. The molybdenum-free
effluent is then passed through the Swift Lectro-Clear heavy metal
removal and cyanide destruction system. Lime is added to
precipitate the heavy metals, and sodium hypochlorite is added to
oxidize the cyanide. The suspension is then passed into a
electrocoagulation cell where oxygen and hydrogen gas are
generated by electrolysis. Aided by an added polymer flocculent,
the "micro bubbles", produced by electrolysis, cause the
precipitated metals to float to the surface of the
electrofloatation cell; the metals are then skimmed off for
disposal. A high-flow-rate sand filter was added to ensure the
quality of the effluent.

Removal of the metals and cyanide was as follows:

1) Copper, from 0.19 to 0.013 milligrams (mg) per liter, 93.2-percent removal;

2) Iron, from 4.3 to 0.10 mg per liter, 99.9-percent removal;

3) Manganese, from 6.4 to 0.10 mg per liter, 99.9-percent removal;

4) Zinc, from .29 to 0.022 mg per liter, 93.4-percent removal;

5) Molybdenum, from 5.9 to 0.6 mg per liter, 90.3-percent removal;

6) Cyanide, from .28 to 0.01 mg per liter, 96.4-percent removal.

PROCESS DIMENSIONS

WASTES GENERATED	Quantity requiring treatment reduced
PRODUCTIVITY	No change
PRODUCT QUALITY	No change
DOWN TIME	No change
FACILITIES	Information not provided
EQUIPMENT	Investment required, details not given
RAW MATERIALS	Conservation of raw materials
WATER	Information not provided
ENERGY	Information not provided
WASTE DISPOSAL	Information not provided
POLLUTION CONTROL	Cost reduced
PERSONNEL/MAINTENANCE	Information not provided
NET BENEFITS	Enhanced conservation

For more information, refer to "Development of Waste Water Treatment at the Climax Mine," by R. D. Gott, in Mining Congress Journal, April 1978, pages 28-34.

Case Study 3

TYPE OF INDUSTRY: Food Canning

NAME OF FIRM: California Canners and Growers (Pardee
                    Engineering, consulting firm, James Dana
                    415/845-4516)

LOCATION: San Jose, California

SIC: 2033

```
*  ---------------------------------------------------------------  *
*                                                                   *
*                         MODIFICATION                              *
*                                                                   *
*            RECYCLING OF WASTE STEAM IN CANNING COOKERS            *
*                                                                   *
*  ---------------------------------------------------------------  *
```

Abstract

      California Canners and Growers (CCG) previously operated
their atmospheric cookers one-half to two-thirds full of water.
Any steam that did not condense in the water escaped through the
can in-feed and discharge ports and was wasted. To utilize this
wasted food-processing energy, CCG, aided by Pardee Engineering,
devised a modification to the existing equipment design which
allowed retention of steam in the cooker. They also removed the
water in the cooker so that the steam condenses directly on the
cans. This has resulted in considerable energy, water, and labor
savings.

ANNUAL COST SAVINGS: $45,000

PAYBACK PERIOD: 4 months

TIME TO IMPLEMENT: 1 day

TRANSFERABILITY: Applicable to food canning operations
                            equivalent to the one described here.

ENVIRONMENTAL AND
HEALTH BENEFITS: Conservation of energy and water;
                           improved health and safety conditions
                           for plant personnel.

Project Description

California Canners and Growers previously operated their atmospheric cookers one-half to two-thirds full of water. Steam, rising from spargers in the bottom of the cooker, condensed in the water. Steam escaped through the can input and discharge ports, resulting in a substantial waste of energy. To avoid condensation from this waste, each port had a steam exhaust stack which projected through the roof. The steam energy, as it flowed to the roof, heated the inside of the cannery building, much to the discomfort of the workers; it was also wasted to the atmosphere.

To save this energy, a steam-jet vacuum compressor was added; it pulls vapors from the can discharge hood and reinjects the compressed vapors, through spargers, into the cooker. Vapor exhaust hoods containing tunnel seals were installed at both can ports to retain steam in the cookers as the cans enter and leave the cooker.

During the testing of these modifications, it was found that steam, bubbling upward from the spargers through the water, had less opportunity to collect and condense on can surfaces than in a pure steam environment. Steam that did not condense in the water or on the cans was vented, and lost its latent heat of evaporation to the atmosphere and was wasted. Consequently, the water was removed, and the cooker became a waterless cooker. This modification provides energy savings of $5,000 per cooker per 1,000-hour season. Thirty-six hundred gallons of water are saved per cooker, and start-up takes only 8 minutes instead of 70 minutes.

Treatment of the cooking water, to keep scale and solid organics off the cans, is no longer necessary; the cans now remain bright and shiny in the steamy environment.

38

PROCESS DIMENSIONS

WASTES GENERATED	3,600 gallons of water conserved per cooker
PRODUCTIVITY	Improved
PRODUCT QUALITY	Improved
DOWN TIME	Decreased 50 percent
FACILITIES	$500 to $1,000 per cooker
EQUIPMENT	$800 to $1,500 per cooker
RAW MATERIALS	Information not provided
WATER	33,400 gallons saved per year
ENERGY	$45,000 saved per 1,000-hour season
WASTE DISPOSAL	Reduction
POLLUTION CONTROL	No change
PERSONNEL/MAINTENANCE	Reduction in start-up time saves substantial employee time
NET BENEFITS	Substantial reduction in operating costs, and improvement in worker comfort

For more information, refer to "Waterless Atmospheric Cookers Save $5,000 Per Year in Steam Costs" by Walter Doucett, Victor Merlino, James E. Philpot, and Kark Robe, in _Food Processing_, November 1982.

Case Study 4

TYPE OF INDUSTRY: Vegetable Processing

NAME OF FIRM: Dunbar Foods
                    (Wayne Dail 919/892-3175)

LOCATION: Dunn, North Carolina

SIC: 2033

```
*  ┌─────────────────────────────────────────────────┐  *
*  │                 MODIFICATION                      │  *
*  │                                                   │  *
*  │    LAND APPLICATION OF FOOD PROCESSING WASTES     │  *
*  │                                                   │  *
*  └─────────────────────────────────────────────────┘  *
```

Abstract

    Dunbar Foods processes and cans peppers and sweet potatoes.
They have developed a spray-irrigation land application system for
their food-processing wastes. Process water containing
vegetable wastes is reused to irrigate and fertilize grassland
owned by the company.

ANNUAL COST SAVINGS: Information not provided.

PAYBACK PERIOD: Information not provided.

TIME TO IMPLEMENT: Information not provided.

TRANSFERABILITY: This modification would be suitable
                            for any firm producing large quantities
                            of high biological oxygen demand (BOD)
                            non-toxic wastewater. Because Dunbar
                            Foods was constrained by the capacity of
                            their municipal waste treatment
                            facility, this innovative method was
                            cost-effective because it enabled them
                            to expand their operation. In areas
                            where the publicly-owned treatment plant
                            can handle the required loading, firms
                            might find it more cost-effective to use
                            the municipal wastewater treatment
                            facility and pay a sewer use-charge. In

41

either case, additional activities might
include a review of those processes
which generate wastewater to find ways
to reduce the waste in strength or in
volume.

ENVIRONMENTAL AND
HEALTH BENEFITS: Excessive organic loading to the
municipal treatment plant is prevented;
reduces risk of inefficient municipal
treatment of wastewater and consequent
pollution load to the receiving stream.
Monitoring revealed no deleterious
changes in nearby groundwater quality,
in availability of oxygen to soil
microbes, or in runoff water quality due
to landspreading activities.
Conservation of water, land, and
material resources: less municipal
wastewater sludge to manage, less
fertilizer and potable water required
for crop production. Maintenance of a
monitoring program is required to ensure
consistent system performance.

Project Description

Dunbar Foods processes diced peppers and sweet potatoes for
sale to market. They became the first industry in North Carolina
to employ spray irrigation for the land application of treated
wastewater. The system was originally developed in 1971, by the
company's engineer to allow expansion of the firm's capacity
without overtaxing the municipal wastewater treatment facility.
Dunbar Foods expanded again in 1982; because of the spray
irrigation system, they have not required the use of the municipal
treatment plant. Process water flow from the firm's cleaning and
canning operations currently totals about 7.5 million gallons per
day. All of this flow is screened for solids removal and is land-
applied to the 85 acres of grassland owned and maintained by
Dunbar. Because Dunbar's operation is seasonal, the majority of
their activity occurs from July through December. The fields are
periodically mowed to maintain a maximum 9-inch growing height.

Dunbar's innovative system for handling high-BOD industrial
wastewater has been studied by the North Carolina Department of
Natural Resources and Community Development. Over a three-year
period of observation and testing, they found no evidence of
groundwater contamination, odor, runoff, or of any other problems.

After ten years of operation, there has been no indication of anaerobic action in the soil, and no change in the quality of the groundwater in numerous observations from test wells, both within the irrigated area, and on its perimeter.

Dunbar also recycles process water used for cleaning. Recycled water is screened and chlorinated before reuse. Final rinse water and water for canning are not reused.

PROCESS DIMENSIONS

==

WASTES GENERATED	Eliminated 7.5 million gallons per day of loading to municipal system
PRODUCTIVITY	No change
PRODUCT QUALITY	No change
DOWN TIME	No change
FACILITIES	Information not provided
EQUIPMENT	Information not provided
RAW MATERIALS	No change
WATER	Information not provided
ENERGY	Information not provided
WASTE DISPOSAL	Information not provided
POLLUTION CONTROL	Information not provided
PERSONNEL/MAINTENANCE	Information not provided
NET BENEFITS	Permitted plant expansions otherwise precluded by limited capacity of municipal wastewater treatment facility

==

Case Study 5

TYPE OF INDUSTRY: Brewing

NAME OF FIRM: Miller Brewing Company
(Kenneth R. Croft 919/627-2100)

LOCATION: 863 East Meadow Road
P.O. Box 3327
Eden, North Carolina 27288

SIC: 2082

```
*  ┌──────────────────────────────────────────────────────┐  *
*  │                    MODIFICATION                       │  *
*  │                                                       │  *
*  │          SALE OF TREATED BREWERY WASTES FOR           │  *
*  │            AGRICULTURAL LAND APPLICATION              │  *
*  └──────────────────────────────────────────────────────┘  *
```

Abstract

Food-processing wastes generated during the brewing operation at Miller Brewing Company are treated and made available to farmers for land application. The by-product is used for liming and as a nitrogen supplement. The reuse of this previously landfilled material has eliminated disposal costs for Miller, provided an inexpensive substitute for standard agricultural liming material, and preserved valuable landfill capacity.

ANNUAL COST SAVINGS: $88,000

PAYBACK PERIOD: Not applicable.

TIME TO IMPLEMENT: 1 year

TRANSFERABILITY: The important features of the Miller Brewery wastes are that they are non-toxic, solid, and contain material suitable for agricultural purposes. These characteristics are common to many food-processing wastes; therefore, this modification may have wide application. The feasibility of this

45

modification for other firms would depend on the composition of the waste and the availability of users.

One important market option that should not be overlooked is the use of a waste exchange brokerage for transfer of wastes. Although Miller's wastes are in a solid form, a slurry could also be used for land application. The Miller wastes have distinct value for soil pH adjustment and as a source of usable nitrogen; however, even materials that are limited to improving soil texture would be valuable for agricultural and horticultural purposes.

ENVIRONMENTAL AND
HEALTH BENEFITS: Conservation of land resources.

Project Description

Miller Brewing Company's Eden Brewery wastewater treatment facility is a 5 million gallon per day activated sludge treatment plant. Wastewater is routed to the treatment plant from the brewing, packaging, and utilities areas of the brewery complex. Domestic sewage from the brewery is treated through the City of Eden's municipal treatment system.

During the activated sludge process, daily amounts of excess biomass are removed as waste sludge. The excess biomass is dewatered through high-pressure plate and frame filters after conditioning with lime and related dewatering compounds. After the filtration process is completed, the biomass has the consistency of moist earth and is lime-rich. The company generates approximately 13,500 tons of waste sludge each year (wet weight). Until recently, this material has been buried in the Rockingham landfill, costing the firm $88,000 per year in hauling and landfill fees.

Miller Brewing Company had the biomass analyzed, and determined that it could be registered as an agricultural product. They offered the biomass to local agricultural and farm operations as a "by-product liming material with nitrogen" for land application. Seven and one-half tons of the by-product material are equivalent to 1 ton of standard agricultural lime.

Numerous farmers have asked to purchase this material for application to cropland; Miller Brewing Company is providing this desired material to local farm and agricultural operations at no cost. This method of pollution control creates a beneficial use for a once worthless material, reduces hauling and landfill fee expenses, and preserves landfill space for future use.

Miller Brewing Company also maintains an incentive awards program for employee suggestions which improve operations and reduce costs.

PROCESS DIMENSIONS

===

WASTES GENERATED	Eliminated 13,000 tons of waste sludge per year
PRODUCTIVITY	No change
PRODUCT QUALITY	No change
DOWN TIME	No change
FACILITIES	No change
EQUIPMENT	No change
RAW MATERIALS	No change
WATER	No change
ENERGY	No change
WASTE DISPOSAL	$88,000 saved per year
POLLUTION CONTROL	No change
PERSONNEL/MAINTENANCE	Information not provided
NET BENEFITS	Annual cost savings, provides soils amendment material

===

Case Study 6

TYPE OF INDUSTRY: Hosiery Manufacture

NAME OF FIRM: Hampshire Hosiery (Ellen Knitting Mills)
                    (George Beauvais, Sr. 704/765-9011)

LOCATION: Spruce Pine, North Carolina 28777

SIC: 2251

```
*  ----------------------------------------------------------  *
*                      MODIFICATIONS                           *
*                                                              *
*    1) USE OF A HEAT EXCHANGE SYSTEM TO REDUCE WASTE          *
*       EFFLUENT TEMPERATURES AND ENERGY COSTS                 *
*                                                              *
*    2) REPLACEMENT OF AN OIL-FIRED FURNACE WITH A WOOD-       *
*       FIRED SYSTEM                                           *
*  ----------------------------------------------------------  *
```

Abstract

    Hampshire Hosiery manufactures women's pantyhose. They have
made several modifications to reduce their energy costs and
improve the quality of effluent discharged to the municipal sewer
system. A heat recovery system was installed to capture heat
from spent dye solutions and to preheat subsequent dye operations.
In addition to energy savings, the exchanger system lowers the
temperature of effluent entering the municipal sewer system from
130 to 70 degrees F. A change in furnace equipment now permits
the recycling of locally generated sawdust as a fuel source. The
plant's oil-fired steam boiler was replaced with a wood-fired
system capable of using sawdust from nearby lumbermills as fuel.
The new boiler operation saves the firm $200,000 per year in fuel
costs.

ANNUAL COST SAVINGS: $225,000

PAYBACK PERIOD: 2 years (heat exchanger)
                          4 years (wood-fired boiler)

TIME TO IMPLEMENT: 3 years

TRANSFERABILITY: Using a heat exchanger to capture waste
                             heat is cost-effective if the flow and
                             temperature of the available water are
                             high enough to justify the capital
                             equipment costs. Equipment
                             manufacturers would be a good source for
                             making this determination. Use of a
                             wood-fired furnace may be limited by the
                             availa- bility and .cost of fuel, and by
                             the age and performance of existing
                             equipment.

ENVIRONMENTAL AND
HEALTH BENEFITS: Reduced thermal pollution; reduced air
                             pollution; conservation of energy and
                             material resources.

Project Description

Ellen Knitting Mills, a division of Hampshire Hosiery, was
discharging spent dye-bath water to the municipal sewer system.
The temperature of the discharge water was 130 degrees F, which
caused breakage of the terra cotta sewer piping. In 1981, the
company invested $100,000 in a heat exchange system that lowered
the effluent water temperature to 70 degrees F. Spent dye water
is discharged into a holding vat; from there it enters the
stainless steel heat exchanger. The exchanger is composed of five
pipes which are 30 feet long, and 8 inches in diameter. Inside
each pipe is a bundle of smaller tubes which allow the heat
transfer.

Heat removed from the water is used to preheat incoming feed
water for the dye tubs from 55 degrees F to about 105 degrees F.

The preheating operation saves about 52,000 gallons of fuel
oil per year; the heat exchange system had a payback period of
only two years.

The company made a second, major modification which has also
resulted in significant energy savings and in the reuse of waste
materials. In 1982, Ellen Mills purchased and installed a sawdust
fired steam generating plant for approximately $800,000. The unit
consists of a 12-ton receiving hopper, a conveyor transport
system, a hogger, a 200-ton storage silo, a Dutch oven combustion
chamber, an auxiliary oil burner, and a 400-horsepower boiler.

In the vicinity of Spruce Pine, North Carolina, there are
several sawmills with a ready supply of waste sawdust. The
sawdust is trucked to Ellen Mills, where it is dumped into the

receiving hopper; passed by conveyor belt, under a magnet to remove metals; and loaded onto a hogger, where it is ground to a uniform size. The sawdust is then lifted, by a bucket elevator, to the top of the storage silo which holds an eight-day supply of fuel.

Sawdust exits the silo from the bottom and is collected in a metering bin, which dispenses it to a screw conveyor leading to the boiler. The metering bin has sensing devices to monitor the steam pressure in the boiler; it releases sawdust, as needed, to maintain a constant pressure. The measured charge of sawdust is blown into the Dutch oven where combustion takes place. The heat produced by this combustion is introduced into the boiler where the steam is generated.

The exhaust gases pass through a series of cyclone separators to remove any particles, and then into a smokestack. At this point, a heat exchanger transfers heat from the exhaust and reinjects it under the grate in the combustion chamber. Since Ellen Mills has ample capacity in their dye-house heat exchanger, they are using the preheated water as their boiler makeup water. This results in greater energy efficiency for the system.

The new boiler system has saved 300,000 gallons of fuel oil per year. The sawdust fuel costs approximately $12.00 per ton; the company reports a 66-percent savings in fuel costs. They estimate a payback period of only four years for the new system. The new boiler is also a much cleaner system, since it is equipped with state-of-the-art pollution control devices. Cyclone separators scrub the vapors and prevent the particulate "fallout" that was present with the oil-fired steam generator.

PROCESS DIMENSIONS

==

WASTES GENERATED	Reduced temperature of waste effluent; reduced particulate air pollutants
PRODUCTIVITY	No change
PRODUCT QUALITY	No change
DOWN TIME	No change
FACILITIES	No change
EQUIPMENT	$100,000 one-time cost of heat exchanger; $800,000 one-time cost of wood fired boiler
RAW MATERIALS	No change
WATER	No change
ENERGY	$225,000 saved per year
WASTE DISPOSAL	No change
POLLUTION CONTROL	No change
PERSONNEL/MAINTENANCE	No change
NET BENEFITS	Annual cost savings, reduced air and thermal pollution, relief from regulatory concerns

==

Case Study 7

TYPE OF INDUSTRY: Furniture Manufacture

NAME OF FIRM: United Globe Corporation
                   (Wayne Kennedy 704-246-5961)

LOCATION: Lexington, North Carolina

SIC: 251

```
*   ------------------------------------------------------------   *
*                                                                  *
*                         MODIFICATION                             *
*                                                                  *
*       INCINERATION OF FURNITURE MANUFACTURING WASTES             *
*                FOR ENERGY AND HEAT RECOVERY                      *
*                                                                  *
    ------------------------------------------------------------
```

Abstract

Two major waste streams, generated during United Globe's furniture manufacture, are incinerated to produce steam and heat. Solids, in the process water wastes from the ragwash and spraybooth are thickened using dissolved air flotation. The resulting sludge is mixed with woodchips, dried, and fed, with semi-solid and liquid finishing wastes, to the primary combustion chamber of a two-stage incineration system. High BTU solvent wastes are fed to the secondary combustion chamber. The steam produced is used for space heating, ragwashing/drying operations, and the sludge dryer. Use of the incinerator has reduced United Globe's waste volume by 95 percent, and the ash produced is not hazardous. Discharge loading to the municipal sewer system has decreased dramatically, as have hazardous waste transport and disposal costs.

ANNUAL COST SAVINGS: $905,000

PAYBACK PERIOD: Less than 3 years

TIME TO IMPLEMENT: 1 year

TRANSFERABILITY: The modifications made by United
                        Globe require a waste fuel source of
                        sufficient volume and BTU-content to
                        offset the capital cost of the inciner-

ator unit. The wastes must also be
free of halogens. Firms, smaller than
United Globe, that cannot justify the
cost of an in-house unit, can minimize
their disposal costs by selling their
wastes for off-site incineration. A
cooperative arrangement among several
small firms might also allow the
purchase of an incinerator for the
group.

ENVIRONMENTAL AND
HEALTH BENEFITS: Conservation of energy and material
resources; reduced threat of surface- or
groundwater contamination by landfill
leachate; reduced threat of surface-
water contamination through inadequate
municipal treatment of furniture
manufacturing wastes; reduced threat of
accident during hazardous waste
transport.

Project Description

In October 1984, Burlington Industries sold their furniture
division to a private owner; the division was renamed United Globe
Corporation. United Globe reports herein on a significant waste
reduction modification which was implemented before they assumed
ownership, but which continues to provide benefits to the on-going
operation.

United Globe Corporation had two major sources of waste:
ragwash and spraybooth wastewater, and semi-solid and liquid
finishing wastes. The former were discharged to the City of
Lexington's wastewater treatment plant; the latter were drummed
and sent to a landfill in South Carolina. Some of the high-BTU
liquid wastes were sent off-site for incineration and energy
recovery.

The ragwash and spraybooth wastes are generated at three of
the company's furniture plants. They contain oils, finish
residues, and solids. The City of Lexington requested that these
wastes be pretreated before release into the city's sewer system.
The semi-solid and liquid wastes are generated at six of
Burlington's furniture finishing operations; they include up to
4,000 gallons per year of waste solvent. Much of the solvent
comes from furniture stripping. It is a mixture of toluene,
xylene, acetone, ethanol, butanol, isopropyl alcohol, naphtha,
methyl ethyl ketone, and esters. These wastes are usually

contaminated with stains, fillers, glazes, and nitrocellulose. A small amount of waste solvent comes from equipment clean-out and from contaminated batches of finishing material.

In order to avoid the high costs of hazardous waste storage, transport, and disposal, the company sought an alternative method of waste management. Plant and corporate personnel worked on a general solution, then consulted with equipment vendors before selecting particular pieces of equipment. A central facility was constructed to house an incinerator and dewatering equipment. Total cost for the building and equipment was $1.5 million.

The company selected a small incinerator boiler package, manufactured by Consumat, which could burn both solid and liquid waste to produce steam for use as heat, and for ragwash operations. The incinerator operation called for plant wastes to be segregated into four groups: solids, heavy liquids, sludges, and solvents. A central treatment system was also developed to separate the solid materials from the company's ragwash and spraybooth wastewater.

As the system currently operates, spraybooth and ragwash wastewater passes through an oil/water separator, and then into a mixing tank, where it is dosed with acid and polymer to promote formation of precipitate. The precipitate is then removed in a Baker dissolved air flotation unit. The resulting sludge is mixed with woodchips and other high-moisture sludge, and is further dewatered in a Holoflite sludge dryer. The dried sludges are then burned in the lower chamber of the incinerator. Drums of low-moisture-content sludges and various plant wastes are loaded, by a skip loader, to a ram feeder for the lower chamber.

Waste solvents are burned in the upper chamber, which uses No. 2 fuel oil as its primary fuel. The upper chamber runs at a temperature of 1,800 degrees F. The heat from the incinerator fires a boiler that makes steam used to wash and dry rags. In winter, excess heat is used to supplement the plant's space heat. The incinerator's non-hazardous ash is sent to the county landfill. Since incinerator exhausts are well within North Carolina emission standards, no scrubber is needed.

Two well-trained people operate the incinerator, monitor its operation, and perform routine preventative maintenance. Spraybooth wastewater and drummed solvent wastes are picked up at the various United Globe plants and transported to a central facility for dewatering and incineration.

The incinerator began full operation in 1982, and burns approximately 1.5 million pounds of waste each year. The company has decreased its hazardous waste disposal requirements by 95 percent and, based on 1981 data, avoided disposal costs of approximately $905,000 per year. The firm also receives spent solvent wastes from smaller furniture companies for incineration at a charge of $29.00 per drum, thus generating additional revenue. The payback period for the incinerator is expected to be approximately three years, depending on market conditions.

The incinerator system has also provided multiple environmental benefits: the City of Lexington's wastewater treatment facilities have been relieved of a heavy pollution load, no hazardous waste is transported or buried in a landfill, and energy resources are conserved.

PROCESS DIMENSIONS

WASTES GENERATED	95-percent volume reduction of Burlington wastes to 1.5 million pounds per year; hazardous waste (Burlington's plus others') eliminated
PRODUCTIVITY	No change
PRODUCT QUALITY	No change
DOWN TIME	No change
FACILITIES	Included with equipment costs
EQUIPMENT	$1.5 million one-time cost
RAW MATERIALS	No change
WATER	Information not provided
ENERGY	$36,000 fuel cost per year
WASTE DISPOSAL	$905,000 saved per year
POLLUTION CONTROL	Information not provided
PERSONNEL/MAINTENANCE	$60,000 cost per year
NET BENEFITS	Relief from hazardous waste management costs, relief from compliance pressure from municipal treatment facility, annual cost savings

Case Study 8

TYPE OF INDUSTRY: Textile Dyeing and Finishing

NAME OF FIRM: West Point Pepperell
                        (Ken McDowell 919/739-2811)

LOCATION: Lumberton, North Carolina

SIC: 266

```
*  ---------------------------------------------------------  *
*                                                             *
*                        MODIFICATION                         *
*                                                             *
*                    ESTABLISHMENT OF A                       *
*             TOXIC CHEMICALS REVIEW COMMITTEE                *
*                                                             *
```

Abstract

West Point Pepperell is a textile dyeing and finishing plant.
In 1975, the company established a Toxic Chemicals Committee to
review products in use or under consideration, and to evaluate
them according to a wide range of health and environmental impact
criteria. The Committee has proven to be an effective management
initiative which has minimized hazardous waste production and its
accompanying costs and environmental impacts.

ANNUAL COST SAVINGS: Not applicable.

PAYBACK PERIOD: Not applicable.

TIME TO IMPLEMENT: 1 year

TRANSFERABILITY: This management program is a preven-
tive measure, transferable to any firm
making significant use of a number of
different chemicals. Time and money
are invested in preventing pollution
rather than in cleaning it up.

ENVIRONMENTAL AND
HEALTH BENEFITS: Reduced risk to the environment;
                          improved health and safety conditions
                          for plant personnel.

Project Description

West Point Pepperell is the third-largest publicly held textile manufacturer in the nation. They have 22,000 employees at 40 manufacturing facilities located in Georgia, Alabama, North and South Carolina, Florida, Maine, Texas, and Virginia. The establishment of a Toxic Chemicals Committee to evaluate and screen chemicals used in the company's operations, was first proposed by the director of the medical department. It was suggested that an interdisciplinary team be appointed to ensure that chemical safety be a priority in the firm.

In 1975, a committee was assembled which included a medical doctor, an industrial hygienist, three research chemists, a professional engineer, a corporate attorney, a corporate safety officer, a representative from chemical production, and a corporate information specialist. In 1976, the committee began its evaluations. It reviewed both chemicals in use and new products being considered for use.

A number of criteria, in addition to the original considerations of personnel safety and fire hazard potential, were used for evaluation. These included:

o hazardous waste characteristics
  (ignitability, toxicity, corrosivity, reactivity)
o "priority pollutant" status
o availability of safer alternatives
o biodegradability
o heavy metal content
o potential for accumulation in the facility
o potential for release to the environment
o hazard potential when mixed with other chemicals
o proposed manner of use
o ultimate fate of the chemical
o hazard potential to the consumer

Products such as hydrazine and hydrofluoric acid were eliminated from use due to limited ventilation.

New products considered and rejected by the Committee included dichlorobenzidine dyes, because of their benzidine base, and a chloride catalyst for resins, because of its association with bis-(chloromethyl) ether. Another committee recommendation involved the substitution of SAF-T-Kleen, a water-based cleaner, for organic solvent cleaners.

When it was first organized, the Committee scheduled regular quarterly meetings. Part of its function was to establish a set of workable procedures to be followed by each plant prior to any chemical purchases. Once these procedures were in place, much of the Committee's work could be handled in-plant. Currently, all new chemical purchase requests must be approved by the Corporate Research Center's chemist, who follows guidelines set forth by the Committee.

West Point Pepperell is very satisfied with its chemical safety review program. They attribute their extremely low levels of hazardous waste production to the work of the Toxic Chemicals Committee. They believe that controlling chemicals at their point of use, and thus avoiding waste management costs, has helped reduce total costs. They currently generate only a small amount of spent solvent in a dry cleaning operation. The solvent is reclaimed out-of-house by a refining company.

PROCESS DIMENSIONS

===

WASTES GENERATED	Minimized use of toxic and hazardous chemicals
PRODUCTIVITY	No change
PRODUCT QUALITY	No change
DOWN TIME	No change
FACILITIES	No change
EQUIPMENT	No change
RAW MATERIALS	Safer, easier to handle
WATER	No change
ENERGY	No change
WASTE DISPOSAL	Minimized
POLLUTION CONTROL	Minimized
PERSONNEL/MAINTENANCE	Reduced exposure to hazards
NET BENEFITS	Reduced hazardous waste generation, enhanced workplace safety

===

Case Study 9

TYPE OF INDUSTRY: Manufacture and Printing of Specialized
Product Labels

NAME OF FIRM: Rexham Corporation
(William J. Klein, Jr. 919/292-9911)

LOCATION: Greensboro, North Carolina

SIC: 275

```
*  ---------------------------------------------------------  *
*                                                             *
*                        MODIFICATIONS                        *
*                                                             *
*    1)  SUBSTITUTION OF WATER-BORNE INKS FOR SOLVENT-        *
*        BASED INKS                                           *
*                                                             *
*    2)  RECLAMATION OF WASTE INK SOLVENTS BY IN-HOUSE        *
*        DISTILLATION                                         *
*                                                             *
*    3)  USE OF A CLOSED-LOOP WASTEWATER SYSTEM FOR           *
*        HEAT RECOVERY                                        *
*                                                             *
*    4)  SEGREGATION OF SPENT CLEANING SOLVENTS FOR REUSE     *
*                                                             *
*    5)  RECOVERY AND SALE OF VAPORS COLLECTED FROM           *
*        OIL-BASED SOLVENTS                                   *
*                                                             *
*  ---------------------------------------------------------  *
```

Abstract

The Rexham Corporation facility in Greensboro, North Carolina, is involved in the manufacture and printing of specialized product labels. They have made several modifications in their printing operations at various North Carolina facilities, to reduce hazardous waste production. A material substitution, replacing some of their alcohol/acetate inks with a water-borne ink, in some applications, reduced the volume of spent solvent requiring disposal and lowered air emission levels. The company also installed a distillation unit to reclaim n-propyl alcohol from their waste inks. The unit reduced their hazardous waste disposal costs by 74 percent and decreased their raw material costs by 16 percent.

ANNUAL COST SAVINGS: $15,000 (distillation system)

PAYBACK PERIOD: Less than 1 year (distillation
                              system)

TIME TO IMPLEMENT: 3 months (distillation system)

TRANSFERABILITY: The suitability of water-borne inks
                              would depend on the material receiving
                              print, and on the gloss and quality re-
                              quired. Also, some printing speed must
                              be sacrificed. Water-based ink sup-
                              pliers may provide assistance in
                              determining optimum conditions for
                              various applications.

                              In-house solvent recovery by
                              distillation is an excellent invest-
                              ment for many firms using volumes of
                              solvent sufficient to justify the
                              equipment costs. When transportation,
                              insurance, and service fees are includ-
                              ed, in-house systems are often quite
                              competitive with off-site recycling
                              arrangements. In-house recovery also
                              eliminates the risk of spills and
                              accidents in transit.

ENVIRONMENTAL AND
HEALTH BENEFITS: Reduced threat to air and groundwater
                              quality; conservation of resources;
                              improved health and safety conditions
                              for plant personnel.

Project Description

Rexham Corporation in Greensboro, North Carolina manufactures and prints product labels. Several of the waste streams from their processes are hazardous and include spent solvents. Because the company's management felt a strong commitment to comply with both the spirit and the technical requirements of the RCRA, the firm began to look, in a comprehensive way, for opportunities to limit the amount of waste they produced. The Greensboro plant made two major modifications to achieve these goals.

Rexham investigated the substitution of water-based inks for the traditional alcohol/acetate-based inks used in their flexographic printing process. They found there were several technical problems with the substitution, but, for many applications, operator retraining was sufficient to achieve excellent product quality. The water-based ink operation reduced

63

both spent solvent volume and air emissions of solvents. The substitution did result in decreased printing speed, and cannot be used if gloss is required in the final product. However, Rexham is optimistic that the ink companies will overcome some of these problems in the near future.

To reduce the volume of waste solvent generated from press and roller cleanup, Rexham installed a Cardinal distillation unit. The unit receives a solvent blend consisting mainly of normal propyl alcohol and a small percentage of normal propyl acetate. The unit was placed in a corner of a materials storage room with a special block wall built around it to ensure that all materials being redistilled were properly contained. The unit boils the solvent out of a 55-gallon drum of press washup and recondenses the clean solvent vapor into another drum for reuse. Total solvent reclamation is not possible because the 55-gallon drum of press washup is heated by a band heater wrapped around the bottom of the drum. Even heat distribution inside the drum cannot be achieved. The press washup at the outside edge of the drum is hotter than that at the center. The residue from this unit is a flammable hazardous waste.

Total cost of the distillation unit and the structural modifications was approximately $16,000. The distillation unit recovers 85 percent of the solvent in the waste stream, resulting in a savings of $15,000 per year in virgin solvent costs, and in a $22,800 savings in hazardous waste disposal costs. After implementation of the solvent recovery process, Rexham sends 72 drums of waste per year to the hazardous waste landfill as opposed to the previous 300 to 350 drums per year. The reclaimed alcohol/acetate mixture is reblended with virgin materials to provide the required composition for new ink batches and for cleanup. The still bottoms are transported off-site for incineration, at a cost of approximately $98.50 per drum.

Rexham plans to add another distillation unit to recover the remaining 15 percent of the solvents in the sludge wastes. They are investigating the purchase of a "second generation" solvent reclamation unit distributed by Activation, Inc., of Charlotte, North Carolina. This unit is very similar to the Cardinal unit except that instead of evaporating the solvent out of a 55-gallon drum, a special holding tank is used. The special holding tank is shallow and is part of the distillation unit. Heat coils run through the sides and the bottom of the tank. This greatly increases the heated surface area and allows an almost total recovery of solvent from the press washup. The residue resulting from this second distillation unit should be non-hazardous waste that can be sent to a sanitary landfill. The new unit is especially appealing because it will allow the Greensboro plant to become declassified as a hazardous waste generator. The company believes relief from the training, verification, reporting, and inspection costs of hazardous waste management is, in itself, a great savings. Once the Activation unit is installed, Rexham will use a two stage method of solvent reclamation. They will utilize the Cardinal unit as the first reclaiming stage. This arrangement will allow the purchase of a smaller Acitivation unit, and will cut the capital investment required by two-thirds.

The Greensboro plant also employs a closed-loop wastewater system. Heat is extracted from an equipment cooling process and is used to heat the firm's warehouse.

PROCESS DIMENSIONS

WASTES GENERATED	Reduced volume of spent solvent by 13,000 gallons per year due to distillation unit
PRODUCTIVITY	Slight loss of printing speed with water-based ink
PRODUCT QUALITY	Sacrifice of some quality with water-based ink
DOWN TIME	Information not provided
FACILITIES	$4,000 one-time cost for distillation equipment
EQUIPMENT	$12,000 one-time cost for distillation operation
RAW MATERIALS	$15,000 saved per year from distillation operation
WATER	Information not provided
ENERGY	Information not provided
WASTE DISPOSAL	$22,800 saved per year from distillation operation
POLLUTION CONTROL	Information not provided
PERSONNEL/MAINTENANCE	Slight increase in cost due to distillation operation
NET BENEFITS	Annual cost savings, reduced RCRA liability and other regulatory concerns

Case Study 10

TYPE OF INDUSTRY: Manufacture of Solvents and Chemical
                    Additives

NAME OF FIRM: Exxon Chemical America (201/474-0100)

LOCATION: Linden, New Jersey

SIC: 2869

```
*--------------------------------------------------------------*
*                       MODIFICATIONS                          *
*                                                              *
*   1)  ADDITION OF FLOATING ROOFS AND CONSERVATION VENTS TO   *
*       RETAIN SOLVENTS                                        *
*                                                              *
*   2)  IMPLEMENTATION OF AN EMPLOYEE STEWARDSHIP PROGRAM      *
*--------------------------------------------------------------*
```

Abstract

Exxon Chemical installed floating roofs over its tanks of
volatile solvents. This has prevented the loss of more than $1.5
million worth of solvents. Concurrently, they installed
conservation vents on the plant's large solvent tanks; the vents
reduced organic air emissions from the tanks by 30 percent to 75
percent. These two equipment design modifications have, in
addition to reducing air pollution, conserved a significant
amount of solvent resource. A management initiative created a
stewardship program to reduce fluctuations in the level of
organic contaminants in the company's wastewater. Sampling
stations were established to monitor concentrations of organic
chemicals. As a result, operators can trace any fluctuations to
their origin and quickly remedy the problem. The information
gained from the monitoring is also used to charge each process
with a portion of the costs of the wastewater treatment system,
and to set targets for waste reduction for each process. This
program resulted in a 75-percent reduction in the quantity of
organic wastes entering the wastewater treatment plant.

ANNUAL COST SAVINGS: $150,000
                             (from floating roofs, average annual
                             savings based on 1983 solvent prices)

PAYBACK PERIOD: Less than 1 year

TIME TO IMPLEMENT: Information not provided.

TRANSFERABILITY: Applicable to industries holding large
                         amounts of solvents and to any firm
                         that has not implemented a waste
                         monitoring/ auditing system.

ENVIRONMENTAL AND
HEALTH BENEFITS: Conservation of solvents; reduction in
                         air pollution by solvents and any
                         associated risks.

Project Description

In 1976, New Jersey's air pollution control regulations made
floating roofs and conservation vents mandatory for the storage
of many types of organic chemicals. In 1975, Exxon, anticipating
these state regulations, installed eleven floating roofs on tanks
holding organics. Upon realizing the raw material savings, they
added five more floating roofs. The floating roof is a cover
that rests on the surface of the fluid, preventing it from
vaporizing; it costs between $500 and $1,000 per tank. The
conservation vents release vapors less freely than do the
standard vents.

Exxon Chemical instituted a stewardship, or waste audit
program which established a series of wastewater sampling
stations to monitor concentrations of organic chemicals. As a
result, in the event of an upset, the plant operators are alerted
to the problem, and can quickly trace and deal with it at its
source. The information gathered by the sampling stations is
also used to charge each process with the cost of operating the
wastewater treatment system that is proportional to the amount of
waste generated by that process. With this information,
management can set waste reduction goals for particular
production areas.

68

PROCESS DIMENSIONS

==

WASTES GENERATED	Total reduction of 1.5 million pounds per year
PRODUCTIVITY	Storage utility enhanced
PRODUCT QUALITY	No change
DOWN TIME	Not applicable
FACILITIES	No new facilities required
EQUIPMENT	$5,000 to $13,000 cost per tank
RAW MATERIALS	Conserved
WATER	Not applicable
ENERGY	Information not provided
WASTE DISPOSAL	Not applicable
POLLUTION CONTROL	Economically cost-effective
PERSONNEL/MAINTENANCE	Information not provided
NET BENEFITS	Conservation savings, conformance with air quality regulations

==

The information for this case study was provided by Mr. David Sarokin and Dr. Warren Muir of Inform, Inc., 381 Park Avenue South, New York, NY 10014. It was compiled as part of their research for a study entitled "Hazardous Waste Reduction in the Organic Chemical Industry."

Case Study 11

TYPE OF INDUSTRY: Nuclear Fuel/Chemical Manufacturing

NAME OF FIRM: Allied Chemical (Edward Shields 201/455-5630)

LOCATION: Metropolis, Illinois

SIC: 2819

```
*  ┌─────────────────────────────────────────────────┐  *
*  │                 MODIFICATION                      │  *
*  │                                                   │  *
*  │         RECYCLING OF FLOURIDE BY-PRODUCT           │  *
*  └─────────────────────────────────────────────────┘  *
```

Abstract

    A $4.5 million investment in recycling equipment is
paying off at a rate of $1 million a year for Allied Corporation.
Allied's Metropolis, Illinois, plant recovers 8,000 tons of their
calcium fluoride annually. The inorganic chemical is then used
as a raw material in anhydrous hydrofluoric acid production at
another facility. Since 1982, over 1,000 cubic yards of hazardous
calcium fluoride wastes have been recycled monthly, saving about
$300,000 a year in disposal and storage costs alone. The process
also recovers about 1,000 tons of lime annually.

ANNUAL COST SAVINGS: $1,000,000

PAYBACK PERIOD: 4.5 years

TIME TO IMPLEMENT: 4 years

TRANSFERABILITY: Applicable to many industries which
                             precipitate wastes from wastewater
                             unmindful of opportunities for the
                             reuse of the substances in other
                             processes. Requires waste exchange
                             handling of residues, unless a known
                             facility is available to receive the
                             material as a resource.

71

ENVIRONMENTAL AND
HEALTH BENEFITS:

Reduced threat to surface- and groundwater quality; conservation of resources; improved health and safety conditions for development downstream from the plant.

Project Description

Allied Chemical, in the process of manufacturing uranium hexafluoride and sulfur hexafluoride, generates a number of waste streams containing soluble fluoride ions. The waste streams are treated with lime to precipitate the fluorides as insoluble calcium fluoride. The mixture was stored in large surface impoundments to give the insoluble calcium fluoride time to settle and form a sludge. The resultant supernatant and precipitate were found to be hazardous wastes under the RCRA because of their high pH.

Due to the large volume of waste being generated and the limited availability of land suitable for the construction of more impoundments, Allied was pushed to seek alternative strategies for handling their wastes; they have since devised a novel method of recycling their residues. The sludge from the lime-treated fluoride waste streams and impoundments was determined to be primarily calcium fluoride, containing 10- to 15-percent unreacted calcium hydroxide, and 10-percent other compounds, primarily sulfates and fluorides. These solids are now processed through a wastewater treatment plant, where they are reacted with a dilute hydrofluoric acid waste stream which neutralizes the sludge and converts the remaining free lime to calcium fluoride. This material has been found to closely resemble fluorspar, the naturally occuring calcium fluoride. The now clean, aqueous waste stream is neutralized and discharged into the river. The calcium fluoride precipitate is collected as a wet solid, and is used at another Allied manufacturing site as a resource for manufacturing anhydrous hydrofluoric acid. The process turns 8,000 tons of waste fluorides a year into a valuable raw material, saving $1 million a year, and eliminating the need for storage or disposal of large quantities of hazardous waste. The initial investment cost for the process modification was $4.3 million.

PROCESS DIMENSIONS

==

WASTES GENERATED	Reduced by 1,000 cubic yards per year
PRODUCTIVITY	Improved
PRODUCT QUALITY	No change
DOWN TIME	No change
FACILITIES	No investment required
EQUIPMENT	$4,300,000
RAW MATERIALS	$50,000 saved per year
WATER	No change
ENERGY	$90,000 cost per year
WASTE DISPOSAL	$300,000 saved per year
POLLUTION CONTROL	$1,000,000 saved per year
PERSONNEL/MAINTENANCE	$410,000 cost per year
NET BENEFITS	Savings through resource conservation and waste reduction

==

For more information, see Chemecology, December 1983/January 1984, p. 5.

Case Study 12

TYPE OF INDUSTRY: Manufacture of Urea and Phenol Resins

NAME OF FIRM: Borden Chemical Company (415/657-4500)

LOCATION: Fremont, California

SIC: 2869

```
*  ————————————————————————————————————————————  *
*                   MODIFICATIONS                  *
*                                                  *
*   1) FILTER RINSE AND SUBSEQUENT REUSE OF PHENOL RESIN  *
*                                                  *
*   2) IMPLEMENTATION OF TWO STAGE TANK RINSING    *
*                                                  *
*   3) MANAGEMENT INITIATIVE IN EMPLOYEE PRACTICES *
*  ————————————————————————————————————————————  *
```

Abstract

Borden Chemical Company reduced the amount of organic phenol resins discharged from their facility by 93 percent, mainly through three process modifications:

1) Phenolic resins are passed through a filtration process that removes the large resinous particles. Previously, the filtered materials were disposed of by rinsing them into the wastewater treatment system. They are now rinsed into large tanks and recycled into the process as raw material.

2) Reactor vessels were filled with water to clean them. This resulted in a large amount of phenolic resin waste that had to be processed through the wastewater treatment system. New equipment operation procedure has reduced the initial rinse to only 5 percent of the previous volume, resulting in a more concentrated solution of phenol resins. This solution is stored and recycled into the process line as a raw material. The reactor vessels are then rinsed a second time. The waste stream from this rinse contains a low concentration of phenolic resins.

3) The Borden Company has made employees aware that small amounts of chemical waste can strongly affect overall waste generation. As a result, management and workers have identified ways of reducing wastes.

ANNUAL COST SAVINGS: Information not provided.

PAYBACK PERIOD: Information not provided.

TIME TO IMPLEMENT: Information not provided.

TRANSFERABILITY: Applicable to industries using
                              batch reactions and one-stage rinses,
                              and to industries that have not
                              implemented a waste monitoring/audit
                              system.

ENVIRONMENTAL AND
HEALTH BENEFITS: Conservation of phenolic resins;
                              reduced risks of leakage from waste
                              storage pond.

Project Description

Borden Chemical Company was using an evaporation pond that removed phenols and other contaminants from their phenolic resin manufacturing line by concentrating them as a sludge for disposal by a waste-handling firm. The evaporation pond had several limitations as a waste management option:

1) The pond filled up faster than expected;

2) The cost of enlarging the pond to accommodate the large amounts of water and sludge was prohibitive;

3) Continued operation of the pond would change Borden's classification to a hazardous waste storage, treatment, and disposal facility. The RCRA requirements for this classification are complex and highly expensive to meet;

4) Reliance on the pond presented the small, but ever-present possibility of a leak which could contaminate groundwater and expose Borden to high remedial costs and long-term liabilities;

5) Borden did not want to continue paying the rapidly rising costs of disposing of the hazardous sludge.

Faced with all these factors, Borden's management looked at ways to decrease the amount of phenols that entered the wastewater system by using two procedures:

1) Plant personnel analyzed the ways in which phenol wastes from the resin operations entered the wastewater treatment system. They monitored wastewater for organic loading in order

to spot unexpected surges, and reviewed staff activities and plant operations to identify the source of the organics;

2)   Plant personnel "walked through" the plant's operations -- from arrival, through each step in the movement of materials, to the final off-site shipment -- identifying all potential or actual sources of waste generation.

From the information gathered, they modified several processes:

1)   During the manufacture of phenolic resins, large particles are removed by filtration.   Previously, the materials were washed from the filters into the wastewater treatment system. Now, they are collected and stored in holding tanks before being recycled through the process line as needed.

2)   Batch reactor vessels were rinsed between batches by filling them with enough water (20,000 gallons) to reach the agitator blades at the top of the tank.   This produced a large amount of wastewater contaminated with the phenolic resin.   The process was modified to a two-rinse system.   Initially, the reactor is rinsed with a small volume of water (500 to 1,000 gallons), producing a wastewater with a high concentration of phenolic resins.   This is stored for reuse.   The tank is then given a full volume rinse, producing a wastewater that has a very dilute concentration of phenols.

3)   The employees were educated about the influence of small amounts of chemical wastes on the overall amount of waste produced.   Management and workers alike identified ways to reduce the production of wastes.   For example, the truck that delivers phenol to the plant was generating waste by letting the phenol, still in the hose after filling the storage tanks, drip into the wastewater treatment system.   The hose is now flushed with water, and the water/phenol mixture passes into the storage tanks pending use.   The result of this and other modifications was a 93-percent reduction of organics entering the wastewater treatment system.

PROCESS DIMENSIONS

WASTES GENERATED	Reduced by 95 percent
PRODUCTIVITY	No change
PRODUCT QUALITY	No change
DOWN TIME	Information not provided
FACILITIES	No investment required
EQUIPMENT	Recovery tank required
RAW MATERIALS	Reduced demand for virgin resource
WATER	Information not provided
ENERGY	Information not provided
WASTE DISPOSAL	Cost significantly reduced
POLLUTION CONTROL	No change
PERSONNEL/MAINTENANCE	Information not provided
NET BENEFITS	Conservation of raw materials, reduced waste disposal costs, and reduced likelihood of environmental pollution

The information for this case study was provided by Mr. David Sarokin and Dr. Warren Muir of Inform, Inc., 381 Park Avenue South, New York, N.Y. 10014. It was compiled as part of their research for a study entitled "Hazardous Waste Reduction in the Organic Chemical Industry".

Case Study 13

TYPE OF INDUSTRY: Nylon Yarn Production and Research

NAME OF FIRM: American Enka Company (John Ray 704/667-7351)

LOCATION: Enka, North Carolina

SIC: 2824

```
*                                                              *
*                    MODIFICATION                             *
*                                                              *
*                  SOLVENT RECYCLING                          *
*                                                              *
```

Abstract

American Enka Company is a nylon yarn production and research facility. They require isopropyl alcohol solvent to produce a polymer film product. They have found it economically profitable and environmentally sound to recycle their waste isopropyl alcohol solvent in-house rather than having it recycled by an outside firm. They purchased a used distillation unit and, with appropriate modifications, American Enka is now saving $90,000 per year. They have also been able to reuse the still bottom residues as an asphalt emulsifier in another product line.

ANNUAL COST SAVINGS: $90,000

PAYBACK PERIOD: 1 month

TIME TO IMPLEMENT: 3 months

TRANSFERABILITY: In-house solvent recovery by
                            distillation can be cost-competitive
                            with the use of off-site vendors.
                            Important variables in determining the
                            suitability of this technology include
                            volume of solvent (to justify
                            equipment costs), contaminants,
                            boiling points, and other chemical
                            characteristics. When trans-
                            portation, liability insurance, and

service fees are figured into the cost of off-site recycling, on-site recovery becomes a vaiable economic alternative.

ENVIRONMENTAL AND
HEALTH BENEFITS: Reduced threat to air and groundwater quality from prior waste disposal; conservation of resources; reduced risks of transportation related accidents and environmental contamination.

Project Description

American Enka Company is a nylon yarn production and research facility. One of their processes uses isopropyl alcohol as a solvent for a fatty amine. The end product is a porous film of polymer. They previously employed an outside firm to distill their waste isopropyl alcohol. They then bought it back for reuse in their production line. Average distillation losses were 15 percent, but losses as high as 40 percent had occurred. Also, due to improper cleaning of the distillation column between runs, the returned isopropyl alcohol was often unusable, due to contamination with Dowtherm, benzene, ethyl benzene, methyl benzene, various chlorinated hydrocarbons, and other organic constituents. Each batch of recycled isopropyl alcohol had to be analyzed for contamination, and, if found unsuitable, required disposal arrangements and fees.

To solve this problem, American Enka purchased a used distillation unit for $7,500, and modified it to redistill the isopropyl alcohol in-house. This resulted in a savings of $90,000 per year, since it is less expensive to distill the alcohol on-site than to contract these services to an outside firm. In addition, the in-house distillation is more efficient, recovering 90 percent of the isopropyl alcohol, as opposed to the outside firm's efficiency of 85 percent. American Enka not only reuses the pure isopropyl alcohol produced by the distillation unit, but they also use the still bottoms as an asphalt emulsifier in another product line. The payback period for this project was approximately one month.

PROCESS DIMENSIONS
(1983 dollars)

==

WASTES GENERATED	Reduced by 10,000 gallons per year
PRODUCTIVITY	No change
PRODUCT QUALITY	No change
DOWN TIME	No change
FACILITIES	No change
EQUIPMENT	$7,500 one-time cost
RAW MATERIALS	$90,000 saved per year
WATER	$81 cost per year
ENERGY	$5,520 cost per year
WASTE DISPOSAL	No change
POLLUTION CONTROL	No change
PERSONNEL/MAINTENANCE	Slight increase in cost
NET BENEFITS	Annual cost savings, improved quality control

==

Case Study 14

TYPE OF INDUSTRY: Pharmaceutical

NAME OF FIRM: Riker Laboratories [Minnesota Mining and
Manufacturing (3M)] (Mike Koeningsberger
612/778-4523)

LOCATION: Northridge, California

SIC: 2834

```
*  -------------------------------------------------------------  *
*                                                                 *
*                       MODIFICATION                              *
*                                                                 *
*     SUBSTITUTION OF WATER-BASED SOLVENT FOR ORGANIC-BASED       *
*              SOLVENT USED IN COATING TABLETS                    *
*                                                                 *
   -------------------------------------------------------------
```

Abstract

Riker Laboratories was coating medicine tablets, using several organic solvents. Consequently, they had the potential for exceeding the air pollution limits set by a new California environmental regulation. A water based solvent was developed to replace the organic solvent, and different spray equipment was installed in order to handle the new solvent. This combination of a resource substitution (of the water-based solvent for the organic solvent) and an equipment design modification saves $15,000 per year in solvent costs, and prevents the production of 24 tons of air pollution per year. Subsequently, the company determined that $180,000 in pollution control equipment, for organic solvent vapors, was no longer required.

ANNUAL COST SAVINGS: $15,000

PAYBACK PERIOD: Less than 1 year

TIME TO IMPLEMENT: Information not provided.

TRANSFERABILITY: Applicable, in many instances, to carriers in coating processes.

ENVIRONMENTAL AND
HEALTH BENEFITS: Elimination of air contamination by
                                solvents; elimination of risks of
                                solvent inhalation by workers and
                                non-workers.

Project Description

     Riker Laboratories, one of 3M's pharmaceutical plants, was
coating medicine tablets with organic-solvent-based solutions.
This method had the potential to exceed air pollution limits set
by a new California environmental regulation. 3M could have
installed pollution control equipment, to capture the solvent
vapors, at a cost of $180,000. Instead, they replaced the
organic solvent coating with a water-based one. Different spray
equipment was installed in the coating machine to accommodate the
new solvent. The pumping mechanism, tubing and control systems
were also modified. Because the switch was from organic solvents
with a higher volatility than water, the heating capacity of the
dryer had to be increased so that the tablets would dry properly.
Special tests proved the stability of the product, the U.S. Food
and Drug Administration approved it in time to meet the effective
date of the new Califoria environmental regulation, and thus
productivity was maintained.

PROCESS DIMENSIONS

===

WASTES GENERATED	Eliminated 24 tons per year
PRODUCTIVITY	No change
PRODUCT QUALITY	No change
DOWN TIME	No change
FACILITIES	No change
EQUIPMENT	Investment required, details unknown
RAW MATERIALS	$15,000 saved per year
WATER	Information not provided
ENERGY	Information not provided
WASTE DISPOSAL	$180,000 cost averted
POLLUTION CONTROL	No change
PERSONNEL/MAINTENANCE	No change
NET BENEFITS	Reduced costs and environmental hazards

===

Case Study 15

TYPE OF INDUSTRY: Cologne, Toilet Water, and Perfume Production

NAME OF FIRM: The Coty Division of Pfizer, Inc.
                    (R.R. Clarke 919/774-8800)

LOCATION: P.O. Box 1026
                    Sanford, North Carolina 27330

SIC: 2844

```
*                                                                    *
*                         MODIFICATION                               *
*                                                                    *
*             HEAT RECOVERY FROM WASTE IGNITABLES                    *
*                                                                    *
```

Abstract

The Coty Division of Pfizer, Inc., has developed a method of incineration/heat recovery from waste organic chemicals. The ignitable hydro-alcoholic wastes are mixed with fuel oil and burned to produce steam and hot water. The company saves over 1,000 gallons of fuel oil a year, and an unspecified amount of natural gas. The cost of hazardous waste disposal has been eliminated.

ANNUAL COST SAVINGS: $2,800

PAYBACK PERIOD: 2 years and 7 months

TIME TO IMPLEMENT: Information not provided.

TRANSFERABILITY: Although the required boiler
                        modifications were specific to this
                        rather unique fuel source, the
                        fundamental concept of heat recovery,
                        using waste materials with fuel
                        potential, has numerous applications.
                        If fuel source and heat requirements
                        cannot be matched internally, a waste
                        exchange contract might make savings
                        possible for two compatible firms.

ENVIRONMENTAL AND
HEALTH BENEFITS: Elimination of threat to groundwater
                               quality; conservation of land and fuel
                               resources; elimination of fire hazard
                               during transport and at the landfill
                               site.

Project Description

The Coty plant has historically had a small percentage of hydro-alcoholic products that become damaged or outdated, and that must eventually be destroyed. Prior to the Resource Conservation and Recovery Act of 1976, bottles of colognes, perfumes, and toilet waters were crushed, and the residual liquids, as well as the crushed containers, were trucked to a landfill for disposal. However, because hydro-alcoholic products are ignitable and constitute a hazardous waste according to the RCRA, an alternative to landfill disposal was sought.

Several management strategies have helped to reduce the quantity of hydro-alcoholic waste. Substantial reductions have been accomplished through strong marketing campaigns, refurbishing of products returned from customers, and sales through employee and outlet stores. For the small percentage of waste hydro-alcoholic products remaining, the company decided to use their heat value, which is approximately two-thirds that of fuel oil.

A unique system for collecting, distributing, and burning the hydro-alcoholics was designed by Coty personnel. With minor modifications to the existing system and hot water boilers, construction of a collection apparatus, and installation of a special feed system, the hydro-alcoholics are burned in a 20-percent mixture with fuel oil. This fuel mixture burns well; for every gallon of hydro-alcoholics consumed, two-thirds of a gallon of fuel oil is conserved.

Coty has been using the system since November 1980, and reports savings of over 1,000 gallons of fuel oil and an unspecified amount of natural gas per year. The initial cost of the project was $7,500, with a payback period of about 2.7 years. There are some minimum operating expenses, but the company is saving about $2,800, annually since it has eliminated all other waste treatment or disposal costs. There are no hazardous air pollutants released from the boiler, and the system has also eliminated the environmental and health risks associated with landfilling ignitable wastes.

PROCESS DIMENSIONS

==

WASTES GENERATED	Hydro-alchoholic wastes eliminated
PRODUCTIVITY	No change
PRODUCT QUALITY	No change
DOWN TIME	No change
FACILITIES	No change
EQUIPMENT	$7,500 one-time cost
RAW MATERIALS	No change
WATER	No change
ENERGY	$2,800 saved per year
WASTE DISPOSAL	$2,800 saved per year
POLLUTION CONTROL	Information not provided
PERSONNEL/MAINTENANCE	Minor cost increase
NET BENEFITS	Relief from RCRA regulations; disposal cost savings

==

Case Study 16

TYPE OF INDUSTRY: Manufacture of Phenol, Aniline, and Related
                   Products

NAME OF FIRM: USS Chemicals (614/532-3420)

LOCATION: Haverhill, Ohio

SIC: 2869

```
*-----------------------------------------------------------------*
*                                                                 *
*                        MODIFICATIONS                            *
*                                                                 *
*             CONTROL OF ORGANIC VAPORS BY:                       *
*                                                                 *
*      1)  ADDITION OF ADSORPTION UNIT AND SCRAP CONDENSER        *
*                                                                 *
*      2)  INSTALLATION OF FLOATING ROOFS ON STORAGE TANKS        *
*                                                                 *
*-----------------------------------------------------------------*
```

Abstract

    USS Chemicals has reduced the organic chemical emissions
from their phenol manufacturing processes through a series of
equipment design modifications:

    1) Adding a resin adsorption system. This reduced
emissions 80 percent and saves 715,000 pounds of cumene per year.
The cumene savings is worth $175,000 per year.

    2) Adding a surplus condenser that returns the emissions
to the process. This saved 400,000 pounds of cumene, worth
$100,000, in its first year of operation.

    3) Adding floating roofs to acetone storage tanks, and a
stripper unit for acetone recovery.

    Modification #2 arose through employee ideas submitted to
USS Chemicals' "Suggestions for Cost Reduction" (SCORE). SCORE
is a management initiative which provides a direct financial
incentive for employees to identify cost-cutting opportunities.
Employees are rewarded a percentage of the money that their
measure saves the company.

ANNUAL COST SAVINGS: $175,000 for modification #1;
                              $100,000 for modification #2.

PAYBACK PERIOD: 30 days for modification #1;
                              18 days for modification #2;
                              no economic data provided for
                              modification #3.

TIME TO IMPLEMENT: Information not provided.

TRANSFERABILITY: Applicable to industries using large
                              amounts of volatile component; and to
                              industries that have not implemented
                              an awards program for employee
                              suggestions.

ENVIRONMENTAL AND
HEALTH BENEFITS: Reduced threat to air quality;
                              conservation of cumene and acetone;
                              decrease in health risks caused by
                              exposure to phenol and acetone
                              vapors.

Project Description

    USS Chemicals manufactures phenol from cumene, and then uses
the phenol to manufacture other chemicals such as acetone, alpha
methyl styrene, bisphenol A, and aniline. Because these
chemicals are volatile, USS Chemicals has done pollution
prevention work to control vapor emissions. They installed resin
adsorption systems to control cumene emissions from phenol
manufacturing. The modifications reduced air emissions by 80
percent, saving 715,000 pounds of cumene worth $175,000. At the
suggestion of one of the plant operators, they installed a $5,000
surplus condenser on one of the open cumene vents. The condenser
returns the cumene directly into the process line. The
modification recovered 400,000 pounds of cumene, worth $100,000,
in its first year of operation.

    The operator's idea was submitted, evaluated, and
implemented through USS Chemicals' "Suggestions for Cost
Reduction" program (SCORE). SCORE provides a direct financial
incentive to plant employees by rewarding them with a percentage
of the money saved due to their ideas. Suggestions are not
restricted to waste reduction. Plant savings can result from
energy conservation, reduced labor costs, process improvements,
and other measures.

    To reduce air emissions and recover acetone, USS Chemicals
uses floating roofs on their storage tanks as well as a stripper
unit. The company did not report the efficiency of these waste
reduction practices, but they were pleased that the practices
reduced waste economically.

89

PROCESS DIMENSIONS

WASTES GENERATED	Air emissions reduced by 80 percent
PRODUCTIVITY	No change
PRODUCT QUALITY	No change
DOWN TIME	No change
FACILITIES	No investment required
EQUIPMENT	Investment required, no details given
RAW MATERIALS	Saved 1,115,000 pounds of cumene ($275,000) per year
WATER	Information not provided
ENERGY	Information not provided
WASTE DISPOSAL	No change
POLLUTION CONTROL	$5,000 for condenser, other cost details not provided
PERSONNEL/MAINTENANCE	Information not provided
NET BENEFITS	Enhanced conservation profitablity

The information for this case study was provided by Mr. David Sarokin and Dr. Warren Muir of Inform, Inc., 381 Park

Avenue South, New York, NY 10014. It was compiled as part of
their research for a study entitled "Hazardous Waste Reduction in
the Organic Chemical Industry."

Case Study 17

TYPE OF INDUSTRY: Production and Sale of Trace Elements to the Fertilizer Industry

NAME OF FIRM: Industrial and Agricultural Chemicals, Inc. (Randall F. Andrews 919/843-2121)

LOCATION: Route 2
Red Springs, North Carolina   28377

SIC: 287

```
*                                                                    *
*                         MODIFICATION                               *
*                                                                    *
*                   SUBSTITUTION OF INDUSTRIAL                       *
*             BY-PRODUCTS FOR VIRGIN RAW MATERIALS                   *
*                                                                    *
```

Abstract

Industrial and Agricultural Chemicals, Inc., (IAC) uses a materials substitution to reduce their raw material costs in the manufacture of trace-element fertilizer formulations. Suitable inorganic chemical wastes are received from firms that would otherwise pay to landfill them. This arrangement constitutes a reuse of by-products which would conventionally be treated as wastes. Three examples are cited describing the utility of this arrangement to both IAC and to the donor of the waste material.

ANNUAL COST SAVINGS: Information not provided.

PAYBACK PERIOD: Not applicable.

TIME TO IMPLEMENT: Not applicable.

TRANSFERABILITY: This case study demonstrates the utility of the waste exchange concept. In all three examples cited, wastes to be landfilled by one firm were salvaged for recoverable material by another firm.

ENVIRONMENTAL AND
HEALTH BENEFITS: Prevention of groundwater contamination
                          risks due to landfilling; conservation
                          of land and material resources

Project Description

Industrial and Agricultural Chemicals, Inc., (IAC) produces
trace-element formulations for sale to the fertilizer industry.
The elements they market include copper, boron, manganese, zinc,
and magnesium. In several instances, they have been able to
reduce their production costs by substituting waste materials from
other firms for virgin raw materials in their own manufacturing
processes. These substitutions include both on-going receipt of
material, such as copper sulfate and magnesium oxide wastes, and
the single-event acceptance of a batch of magnesium chloride which
had become contaminated with copper chloride. In each case, the
waste materials were otherwise destined for landfill disposal.
IAC, therefore, obtains raw materials at no cost, and the firms
donating the material eliminate their disposal costs, which are
often substantial. Furthermore, usable chemicals are not
discarded, which saves land and material resources, and reduces
health and environmental risks.

With the advent of the Piedmont Exchange (a waste exchange),
IAC's opportunities to accept free industrial wastes have
increased significantly. IAC representatives claim that waste
exchanges have improved their firm's profitability by offering
them more opportunities to reduce their raw material costs.

A waste exchange usually functions as an information
clearinghouse. The exchange lists materials which are no longer
desired by a donor, but may be useful to other parties; or
materials desired by a manufacturer which may be obtained cheaply
if generated in another process as a waste.

PROCESS DIMENSIONS

==

WASTES GENERATED	Variable, no change
PRODUCTIVITY	Improved
PRODUCT QUALITY	No change
DOWN TIME	Not applicable
FACILITIES	No change
EQUIPMENT	No change
RAW MATERIALS	Reduced
WATER	No change
ENERGY	No change
WASTE DISPOSAL	Reduced disposal requirements for donors
POLLUTION CONTROL	No change
PERSONNEL/MAINTENANCE	No change
NET BENEFITS	Cost savings, resource conservation

==

Case Study 18

TYPE OF INDUSTRY: Agricultural Chemical Research and Development

NAME OF FIRM: ICI Americas, Inc.
                    (Paul Ekoniak 919/731-5200)

LOCATION: Goldsboro, North Carolina

SIC: 2879

```
******************************************************************
*                                                                *
*                        MODIFICATIONS                           *
*                                                                *
*    1)  ESTABLISHMENT OF AN ENVIRONMENTAL COMPLIANCE COMMITTEE   *
*                                                                *
*               2)  IMPROVED HOUSEKEEPING                         *
*                                                                *
*                 3)  SOLVENT RECOVERY                            *
*                                                                *
******************************************************************
```

Abstract

ICI, Goldsboro, as a research and development facility, does not produce a large amount of waste. However, to manage the waste it does generate, a management initiative was undertaken; in 1981, ICI formed an Environmental Compliance Committee to review issues and activities that affect the environment. The Committee's primary emphasis is early planning to minimize waste production, and second, attempting to recycle or treat the waste, whenever possible, so that there is zero discharge. Measures recommended by the Committee, and then implemented, include housekeeping improvements and the initiation of waste solvent recovery by distillation. The Committee's recommendations saved the company an estimated $37,000 in waste disposal costs in 1984.

ANNUAL COST SAVINGS: Estimated $37,000 in 1984, compared
                        with conditions in 1981.

PAYBACK PERIOD: Immediate.

TIME TO IMPLEMENT: On-going process.

TRANSFERABILITY: This approach to controlling the
                              generation of waste is broadly
                              applicable to many waste generators.
                              A waste audit, good housekeeping, and
                              stringent segregation of waste streams
                              will usually yield a net reduction of
                              wastes requiring treatment. These
                              efforts are valuable first steps in
                              waste recycling and minimization,
                              especially for the many smaller
                              generators of hazardous waste who have
                              recently been brought under the RCRA
                              regulatory framework. Such generators
                              are characteristically less able to
                              make large capital investments for
                              process modifications or recycling
                              equipment.

ENVIRONMENTAL AND
HEALTH BENEFITS: Reduced threat to air and groundwater
                              quality; conservation of resources;
                              improved health and safety conditions
                              for plant personnel.

Project Description

    ICI, Goldsboro, is involved in research and development of
agricultural chemicals.   They generate a variety of waste
streams,   but the total amount of waste produced is not very
great.   The company's wastes are flammable, non-halogenated
solvents and non-flammable, halogenated solvents, with the RCRA
classifications: F002, F003, F005, D001, and U057.

    Waste management activities at ICI have focused on the
evaluation and planning of any newly proposed projects, in order
to minimize or prevent waste generation.   An Environmental
Compliance Committee was formed to oversee these activities, and
to develop and implement procedures to minimize production of
wastes requiring landfilling.   The Committee consists of nine
persons with various types of expertise, and, as a result of
their efforts, ICI reports a significant reduction in hazardous
wastes sent off-site for burial.

    Although there was no previously existing process with which
to make cost savings comparisons, indications are that a
significant reduction, in hazardous wastes sent off-site for
management, has occurred. In the last three to four years, the
ICI group has more than doubled its research work. In the same

period, ICI has essentially reduced the amount of hazardous waste generated by 70 percent.

In 1981, there was no concerted effort to reduce the generation of waste; one-hundred 55-gallon drums were sent off-site for treatment, recycling, or burial (at an average cost of $130 per drum). In 1984, only 60 drums were sent off-site (at an average cost of $266 per drum). Taking into account the doubled output at the labs, and a roughly 100-percent increase in treatment and disposal costs per drum of hazardous wastes over the past three to four years, ICI's waste prevention strategy saved an estimated $37,000 hazardous waste management costs in 1984.

The following are examples of new practices that have been implemented to help reduce the production of waste:

1) Spent chlorinated and non-chlorinated solvents are now segregated to allow off-site distillation and recovery. Since ICI uses approximately 55 gallons of solvent per week, this has resulted in a cost savings for the firm.

2) In large scale laboratories, separate waste containers are designated for non-hazardous and hazardous waste. This results in less hazardous waste by reducing cross-contamination of the non-hazardous waste by hazardous waste.

3) Unused portions of chemicals from the agricultural research fields are returned to the center for reuse or reformulation.

PROCESS DIMENSIONS

==

WASTES GENERATED	Reduced by 70 percent (140 x 55 gallons per year)
PRODUCTIVITY	No change
PRODUCT QUALITY	No change
DOWN TIME	No change
FACILITIES	No change
EQUIPMENT	Minimal
RAW MATERIALS	No change
WATER	No change
ENERGY	No change
WASTE DISPOSAL	Estimated savings of $37,000 in 1984
POLLUTION CONTROL	No change
PERSONNEL/MAINTENANCE	No change
NET BENEFITS	Reduced waste disposal costs and conservation of resources

==

Case Study 19

TYPE OF INDUSTRY: Pesticide Production

NAME OF FIRM: Daly-Herring Company
                  (D. W. Craig 919/527-8001)

LOCATION: P. O. Box 428, Neuse Road
                  Kinston, North Carolina

SIC: 2879

MODIFICATION

SEPARATION OF WASTE DUST STREAMS TO PERMIT REUSE OF DUST

Abstract

Daly-Herring Company altered their dust collection equipment operation so that waste streams, containing organic chemicals from various production areas, are now collected separately rather than mixed in a single baghouse. The collected materials are no longer contaminated by alternate waste streams, and each is recycled back to the process where it was generated. The firm has eliminated over $9,000 in annual disposal costs, and estimates that the recovered material is worth more than $2,000 per year.

ANNUAL COST SAVINGS: $11,638

PAYBACK PERIOD: 10 months

TIME TO IMPLEMENT: Information not provided.

TRANSFERABILITY: This project demonstrates how waste stream separation can reduce waste production and disposal costs. The separation concept can be applied across industries, processes, and materials. Whenever several waste streams are mixed, separation may reduce the total amount of hazardous

material requiring disposal, or
as in this case, the separated waste
stream components can be recycled to
the process that generated them.

ENVIRONMENTAL AND
HEALTH BENEFITS: Reduced threat to soil and groundwater;
                                    conservation of materials; reduced
                                    risk to employees required to handle,
                                    store, and transport waste materials.

Project Description

Daly-Herring Company manufactures pesticides and generates
approximately 45,000 pounds of pesticide dust per year from two
major production systems in the plant. Until recently, the dust
was collected in a single baghouse and was unsuitable for reuse
because it contained a mixture of several pesticides.
Approximately 112 drums of waste dust were landfilled per year,
at a cost of $28 per drum. Transportation, disposal, and labor
costs totalled over $9,000 a year.

In April 1983, the firm replaced the single baghouse
collection system with two separate vacuum-air-baghouse systems
specific to the two production lines. Dust from the respective
production areas is now collected separately in independent
baghouses. The filtered dust is automatically shaken down into a
hopper which augers the dust back to the production area where it
was generated. The returned dust is classified as inert filler
in the final product because the quantitative analytical testing,
which would be required to characterize it as an active
ingredient, is costly. The recycling reduces costs for raw
production material and eliminates waste dust disposal costs.
The payback period for the $9,600 investment in new equipment and
renovations was less than one year.

PROCESS DIMENSIONS

===

WASTES GENERATED	45,000 pounds of waste dust eliminated per year
PRODUCTIVITY	Slight increase
PRODUCT QUALITY	No change
DOWN TIME	No change
FACILITIES	No change
EQUIPMENT	$9,600 one-time cost
RAW MATERIALS	$2,000 saved per year
WATER	No change
ENERGY	No change
WASTE DISPOSAL	$9,000 saved per year
POLLUTION CONTROL	Information not provided
PERSONNEL/MAINTENANCE	No change
NET BENEFITS	Annual cost savings and reduced liability risks

===

Case Study 20

TYPE OF INDUSTRY: Defluorinated Phosphate Manufacture

NAME OF FIRM: Texasgulf

LOCATION: Saltville, Virginia

SIC: 2879

```
*********************************************************************
*                                                                   *
*                        MODIFICATION                               *
*                                                                   *
*      A CLOSED-LOOP, PROCESS WATER RECYCLING SYSTEM WAS            *
*         INSTALLED, REPLACING THE SINGLE-PASS SYSTEM              *
*                                                                   *
*********************************************************************
```

Abstract

Texasgulf manufactures phosphate products. In designing their new defluorinated phosphate plant to meet EPA regulations pertaining to the release of fluorides into the air and water, they decided to implement a closed-loop, process water system to remove inorganic fluorides from the discharge stream. Other alternatives considered included releasing the partially treated wastewater to Saltville's municipal sewage system or discharging it into the nearby Holton river. Benefits of their decision to implement a closed-loop system to recycle process water, in contrast to a single-pass system, include:

    1) A savings of approximately 280,000 gallons of water per day;

    2) A combined annual savings of $930,000 in water purchases and treatment;

    3) Compliance with all effluent guidelines without a NPDES permit, since there is no industrial discharge.

ANNUAL COST SAVINGS: $2,000,000

PAYBACK PERIOD: 1 year

TIME TO IMPLEMENT: 18 months

TRANSFERABILITY: Applicable to industries using water
                             as their process solvent, and to those
                             which have fluorides as an unwanted
                             by-product.

ENVIRONMENTAL
HEALTH BENEFITS: Reduced threat of groundwater, river
                             water, and air contamination from
                             fluorides; conservation of 100 millon
                             gallons of water per year; decreased
                             health risks from fluoride pollution.

Project Description

     To make phosphate animal-feed supplement, Texasgulf combines
phosphate ore with soda ash and phosphoric acid. They then
process it through a kiln mixer for the initial reaction and
partial removal of the fluorides and of some sulphur compounds
present in the phosphate rock. The fluorides exit from the mix
tank as gases, and are conveyed to the mix plant's venturi
scrubber for removal prior to atmospheric discharge. The reacted
rock is then fed to rotary kilns for final drying and
defluorination. It is then cooled, bagged, and shipped out for
sale. The kiln's exit gas (consisting of 15.32-percent carbon
dioxide, 12.36-percent water, 1.95-percent hydrogen fluoride,
0.27-percent sulfur dioxide, and 67.27-percent of various oxides
of nitrogen, and with a mass flow rate of 1,300 tons per day) is
conveyed through a separate dual-stage quencher/scrubber for
pollutant removal prior to atmospheric discharge. The kiln
scrubber's sump-water effluent has a flow rate of 1,725 gallons
per minute at 108 degrees F. The two liquid waste streams from
the scrubbers are pumped to a single lime treating tank, and two
tons of lime per hour are added to the wastewater. The fluorides
precipitate as calcium fluoride, which is separated from the
process water, passed through two gravity thickeners, a filter
press, and is finally buried in the county landfill as a solid.
The clean process water is recycled back to the venturi scrubbers
for reuse, thereby closing the water loop.

PROCESS DIMENSIONS

===

WASTES GENERATED	Reduced by 280,000 gallons per day
PRODUCTIVITY	No change
PRODUCT QUALITY	No change
DOWN TIME	No change
FACILITIES	No change
EQUIPMENT	No information provided
RAW MATERIALS	No change
WATER	$870,000 saved per year
ENERGY	Information not provided
WASTE DISPOSAL	$930,000 saved per year
POLLUTION CONTROL	Lime cost, details not provided
PERSONNEL/MAINTENANCE	No change
NET BENEFITS	Conservation of water resources, cost savings, compliance with Clean Water Act

===

Case Study 21

TYPE OF INDUSTRY: Road Paving and Maintenance

NAME OF FIRM: Nebraska Small Farms Action Group
(Steve Burdic 402/475-3637)

LOCATION: Lincoln, Nebraska

```
*                                                                      *
*                        MODIFICATION                                  *
*                                                                      *
*      SUBSTITUTION OF AN ASPHALT AND CRUMB RUBBER MIXTURE             *
*                  FOR CONVENTIONAL ASPHALT                            *
*                                                                      *
```

Abstract

A study was performed to determine the feasibility of recycling used tires as a secondary resource for roadpaving and other products which could utilize rubber. The pilot-scale project used buffings from tire retreaders as representative of crumb rubber derived from the cryogenic processing of used tires. In the study, a section of road was laid using the crumb rubber in a two-percent mixture with the conventional asphalt surfacing material. Road life was documented to be double that of conventional asphalt, and economic studies indicated that the test material was cost-competitive with concrete.

ANNUAL COST SAVINGS: Not applicable.

PAYBACK PERIOD: Not applicable.

TIME TO IMPLEMENT: Information not provided.

TRANSFERABILITY: Broadly applicable throughout the U.S., and in any location, with a surplus of used tires, which builds or maintains roadways.

ENVIRONMENTAL AND
HEALTH BENEFITS: Could eliminate the problem of used tire disposal, thus conserving a considerable percentage of landfill volume; would reduce the incidence of

105

air pollution from the burning of
tires; would reduce the number of
tires available as breeding grounds
for insects.

Project Description

The Nebraska Small Farms Action Group and the Institute for
Local Self-Reliance received a grant from the Nebraska's
Environmental Control Department to complete feasibility studies
and testing of crumb rubber as an aggregate additive to asphalt
paving, and as a filler/sealer for cracks in state roads.
Initial indications were that the most promising use of the crumb
rubber was as an additive in asphalt roadpaving.

The pilot-scale operation used buffings from retreaders,
to represent actual crumb rubber taken from cryogenically
processed used tires. The quantity of available buffings would
not support a large-scale project. Investment capital of
$800,000 is being sought to build a processing plant to produce
crumb rubber from discarded tires.

The cost of road construction using the rubber additive
increased by an estimated 10 to 15 percent, but the road life is
doubled due to the greater tensile strength imparted by the
rubber. This made the test asphalt favorably cost-competitive
with concrete.

The report concluded that a two-percent mix of rubber in all
the road asphalt projects in Nebraska would use all of the crumb
rubber which could be produced from the state's discarded tires.
It was further deduced that this ratio would also hold true on a
national level. The New Jersey Office of Recycling reports that
New Jersey generates approximately 7 million waste tires per
year, of which 90 percent are either buried in landfills or
illegally dumped.

In Nebraska, legislation has been proposed to revise state
procurement specifications in order to ensure development of this
market for crumb rubber. The U.S. Department of Transportation
guidelines for financing state roadways already allow an
additional 5-percent revenue-sharing benefit to states that
incorporate recycled materials in their highway projects.

PROCESS DIMENSIONS

==

WASTES GENERATED	Information not provided
PRODUCTIVITY	No change
PRODUCT QUALITY	Improved
DOWN TIME	Not applicable
FACILITIES	Investment, details not provided
EQUIPMENT	Investment, details not provided
RAW MATERIALS	Free discarded tires
WATER	Not applicable
ENERGY	Reduced energy required for paving
WASTE DISPOSAL	Common and associated costs eliminated
POLLUTION CONTROL	Eliminates cost of landfilling tires
PERSONNEL/MAINTENANCE	Information not provided
NET BENEFITS	Improved product quality, improved waste management economics

==

Excerpted from The City of Newark's Waste Utilization Study, prepared by Self-Reliance, Inc., for the Department of

Engineering, Newark, New Jersey, written by Jon Huls, Neil
Seldman, Maurice Sampson, Anne Scheinburg, and Trish Ferrand.

Case Study 22

TYPE OF INDUSTRY: Steel Smelting

NAME OF FIRM: Florida Steel Corporation
Charlotte Steel Mill Division
(Dennis E. Tebby 704/596-0361)

LOCATION: Charlotte, North Carolina

SIC: 3312

```
*  ------------------------------------------------  *
*                                                    *
*                  MODIFICATION                      *
*                                                    *
*        SALE OF WASTE DUST FOR ZINC RECOVERY        *
*                                                    *
*  ------------------------------------------------  *
```

Abstract

Florida Steel Corporation sells metal dust, collected from furnace fumes, to a zinc smelter. This practice has reduced their disposal costs by almost half and allows reuse of the metal, rather than disposal.

ANNUAL COST SAVINGS: $129,600

PAYBACK PERIOD: Immediate.

TIME TO IMPLEMENT: Not applicable.

TRANSFERABILITY: Any firm producing significant quantities of metallic dust could pursue opportunities for selling or donating the dust in order to reduce disposal costs and risks. Information pertinent to such an investigation would include composition analysis of the waste, possible alternative uses, and availability of markets. One important market option that should not be overlooked is the use of a waste exchange brokerage.

ENVIRONMENTAL AND
HEALTH BENEFITS: Conservation of land and material
                                resources; reduced threat of heavy
                                metal contamination of surface- and
                                groundwater from landfill leachate.

Project Description

     Florida Steel Corporation produces steel in two electric
arc-furnaces. Scrap steel is the predominant raw material and
includes such items as shredded automobiles, machine shop
turnings, and heavy plate. The fumes emitted from the furnaces
are ducted to baghouses, where the gas is filtered.
Approximately 2,700 tons of waste dust are collected from the
furnace each year. Because electric furnace dust has been
categorized as a hazardous waste by the EPA, the company has been
unable to landfill the waste dust on-site, as had been done in
the past. The waste material was deposited in a hazardous waste
landfill in South Carolina, at a cost of $109 per ton, which
included transportation, disposal fees, and charges imposed by
South Carolina.

     Chemical analyses of the waste dust revealed high
concentrations of zinc. The firm has recently arranged to ship
the furnace dust to a zinc smelter for metal recovery, at a cost
of $61 per ton. This has resulted in significant cost savings to
Florida Steel, elimination of the pollution risks associated with
the landfilling of zinc, and recovery of a useable material --
zinc oxide.

PROCESS DIMENSIONS

WASTES GENERATED	Reduced
PRODUCTIVITY	No change
PRODUCT QUALITY	No change
DOWN TIME	No change
FACILITIES	No change
EQUIPMENT	No change
RAW MATERIALS	No change
WATER	No change
ENERGY	No change
WASTE DISPOSAL	$129,600 saved per year
POLLUTION CONTROL	No change
PERSONNEL/MAINTENANCE	No change
NET BENEFITS	Relief from the high costs of hazardous waste management

Case Study 23

TYPE OF INDUSTRY: Grey and Compacted Graphite Iron Casting

NAME OF FIRM: Waupaca Foundry (Lancy International, consulting firm, William McLay 412/452-9360)

LOCATION: Waupaca, Wisconsin

SIC: 3321

```
************************************************************
*                                                          *
*                    MODIFICATION                          *
*                                                          *
*   CLOSED-LOOP WASTEWATER SYSTEM REPLACED SINGLE-PASS SYSTEM  *
*                                                          *
************************************************************
```

Abstract

Due to planned plant expansion and new governmental regulations, Waupaca Foundry modified the equipment design of their process water treatment system from a single-pass system to two interconnected closed-loops. Through selective separation of the wastewater streams, Lancy International, the consulting firm, was able to design the closed-loop process so that only 25 percent of the plant's total process water had to be treated. This approach minimizes chemical usage in the wastewater treatment, keeps pumping costs down, and reduces water usage.

ANNUAL COST SAVINGS: $20,995

PAYBACK PERIOD: 3.5 years (after expansion)

TIME TO IMPLEMENT: Information not provided.

TRANSFERABILITY: Applicable to industries using water as their process solvent.

ENVIRONMENTAL AND
HEALTH BENEFITS: Reduced threat of ground- and surface-water pollution; conservation of water and wastewater treatment chemicals; decreased health risks from exposure treatment chemicals.

Project Description

Waupaca originally passed the wastewater from two process streams and the blowdown water from their No. 1 casting plant through a holding pond for cooling. The wastewater then passed through a drag tank, to remove suspended solids, before it was allowed to flow into the river. There were various problems associated with this system:

1) The pond tended to fill up rapidly and allowed a popcorn-like residue to flow over the weir and into the river.

2) Heavy rain churned up the contents of the pond and increased the amount of suspended solids in the outfall.

3) The pond required frequent cleaning -- a task which had to be done on premium weekend time while the foundry was not in operation.

Two additional factors influenced their decision:

1) The government issued new pollution control regulations which would require zero discharge into navigational waters.

2) Waupaca wanted to expand its 60-inch-diameter cupola from a capacity of 15 tons to 18 tons per hour and, at a later date, to 30 tons per hour.

Consequently, Waupaca decided to upgrade their wastewater treatement system. They engaged Lancy International to analyze, engineer, and supply the equipment needed to upgrade the system.

A study by Lancy determined that the most economical and environmentally sound approach would be a closed-loop system that could recycle the process water. Lancy also found that, through the segregation of various streams, only about 25 percent of the plant's total process water had to be treated. Consequently, capital costs for the pretreatment facility could be reduced by 50 to 75 percent. As a result of the study, the closed-loop system was installed. It handles the effluent flow from Plant No. 1's cupola operation. The system provides clarified water that is reused in the venturi scrubber. Water from the quencher flows into the drag tank where the large particles settle out.

Particle-laden scrubber water flows into the separator sump. The overflow from the sump is pumped to the drag tank. The underflow from the sump flows into the neutralization/ flocculation tank. Effluent streams, from the drag tank and

113

thickener overflow, are also conducted into the neutralization/ flocculation tank, where caustic soda and polymer are added to precipitate and hasten settling of the metals from the waste stream. The mixture then flows into a standard lamella-type Lancy clarifier where solids are removed as sludge. The water

from the clarifier flows to the wet well and is recycled through the venturi. The sludge is pressed, and the water obtained is recycled back to the neutralization/flocculation tank. The sludge is taken to the foundry's approved landfill.

Strategies that Waupaca's management found beneficial in planning these modifications include:

1) Using the foundry's exact process water chemistry as a basis for designing the least costly equipment system.

2) Following a "start small" strategy to minimize investment in capital equipment, reduce operating costs, and gain operating experience.

3) Providing flexibility in the system design in order to minimize changes in existing production facilities and operations, to allow for production growth, and to accommodate future pollution control regulations.

PROCESS DIMENSIONS
(1980 dollars)

==

WASTES GENERATED	No change
PRODUCTIVITY	No change
PRODUCT QUALITY	No change
DOWN TIME	No change
FACILITIES	No change
EQUIPMENT	$400,000 savings
RAW MATERIALS	No change
WATER	$13,000 saved per year
ENERGY	$744 cost per year
WASTE DISPOSAL	Information not provided
POLLUTION CONTROL	$21,000 saved per year; approximately $0.25 per ton of production, including water and energy
PERSONNEL/MAINTENANCE	$18,000 cost per year
NET BENEFITS	Conservation, minimization of waste management costs

==

For more information, refer to "Planning and Installing a Waste Water Treatment System" by F.M. Degner, in <u>Foundry Management and Technology</u>, June 1983.

Case Study 24

TYPE OF INDUSTRY: Plumbing Products Manufacture

NAME OF FIRM: Stanadyne, Inc., Moen Group
(George McRae 919/258-3341)

LOCATION: 2609 Cox Mill Road
Sanford, North Carolina

SIC: 3432

MODIFICATIONS

1) MATERIAL SUBSTITUTION TO REDUCE SLUDGE VOLUME

2) CHANGE IN EQUIPMENT AND OPERATION TO REDUCE CYANIDE CONCENTRATIONS IN WASTE

3) REVISED RACK DESIGN TO REDUCE DRAGOUT

4) INSTALLATION OF EQUIPMENT TO RECYCLE METALS IN PLATING BATH

5) IMPROVEMENTS IN FILTRATION EQUIPMENT OPERATIONS TO REDUCE WASTE

6) INSTALLATION OF A FILTER PRESS TO REDUCE WASTE SLUDGE VOLUME

7) INSTALLATION OF NEW WATER PURIFICATION SYSTEM TO IMPROVE EFFLUENT QUALITY

8) HOUSEKEEPING IMPROVEMENTS TO REDUCE WASTE

9) INCORPORATION OF POLLUTION PREVENTION CRITERIA IN NEW EQUIPMENT SELECTION

10) DEVELOPMENT OF A PROGRAM TO PROMOTE POLLUTION PREVENTION

Abstract

The Stanadyne Company undertook a comprehensive and systematic review of their electroplating processes in order to reduce or eliminate waste generation. Their efforts have resulted in a broad spectrum of activities which have saved the firm money and minimized pollution. A sulfuric/peroxide materials substitution for chromic acid reduces sludge production from their parts cleaning operation. The company also found they could operate their cyanide copper- and chrome-plating baths at lower concentrations, saving both raw material and sludge-handling costs. Other equipment design and operation changes, which reduced dragout volumes and rack recoating costs, also improved plating efficiency. Tanks were installed, before the rinsing operation, to recycle valuable metals back to the plating baths. A change in the operation of plating-bath filtration equipment operation drastically reduced solution losses, and resulted in a 93-percent cost reduction for filter media. A filter press operation, for sludge dewatering, minimized the cost of sludge transport and disposal. A new wastewater and sand filtration system has improved the treated wastewater quality. Housekeeping improvements led to leak elimination and to a preinspection program which prevents plating of defective parts. Stanadyne also incorporated pollution prevention goals in their recent plant expansion; the new degreaser they selected has significantly lowered solvent costs. Finally, the company has undertaken numerous management initiatives to promote pollution prevention within their firm, among other electroplaters, and among the state's citizens at large.

ANNUAL COST SAVINGS: Information not provided.

PAYBACK PERIOD: Not applicable.

TIME TO IMPLEMENT: Ongoing, 1982 to present.

TRANSFERABILITY: The most important features of the
                           Stanadyne modifications are the commit-
                           ment the firm has made to waste
                           reduction, and the on-going, systematic
                           approach they have used to achieve this
                           goal. Once they identified the two ma-
                           jor sources of pollution -- dragout and
                           spent process solutions -- they focused
                           their efforts on these features of
                           their operation.

                           With respect to specific modifications
                           described, the following comments are
                           offered:

118

o   The success of changes in solution
    makeup will depend on what materials
    compose the plating bath, and on the
    thickness and deposit quality of the
    coating required.

o   Dragout recovery was a low-cost
    modification for Stanadyne. Some
    kind of dragout recovery system can
    usually be used after most heated
    plating baths. However, some
    applications may be limited by a
    build-up of degraded organic
    additives and metallic contaminants,
    or by the heat sensitivity of some
    of the components. Some systems
    might require prohibitive capital or
    energy costs. Use of this
    modification may also require
    additional floorspace to accommodate
    extra tanks.

o   The housekeeping activities and
    filter media changes demonstrate the
    savings that can be achieved by
    consistent monitoring of equipment
    use, and by keeping abreast of new
    product development. The most
    important features of the Stanadyne
    program are the company's commitment
    to pollution reduction, and the
    willingness of management to devote
    personnel time to both in-house and
    public activities aimed at achieving
    this goal.

ENVIRONMENTAL AND
HEALTH BENEFITS: Conservation of material resources;
                        reduced threat of ground- and surface-
                        water contamination; reduced personnel
                        exposure to cyanide.

Project Description

    The Stanadyne plant in Sanford, North Carolina, manufactures
plumbing products, principally the Moen faucet. Some of the
operations that are essential to the production of a marketable,
quality faucet also result in the production of waste material.
These operations include: plating, bright dipping, stripping,

119

and degreasing. The company has adopted a comprehensive and systematic approach to reducing waste generation. They began with the question: "How can pollution be prevented without adversely affecting the electroplating process?"; they proceeded to carefully review their operation with this question in mind.

The plumbing products that Stanadyne manufactures are plated with copper, nickel, and chrome. These finishes are applied to zinc or brass parts to resist corrosion; it was important that the finish quality not be compromised. The company also wanted to avoid any increase in plating defects or contamination of plating baths which could ultimately lead to more, rather than less, waste production.

The modifications reported herein come from all phases of their operation including: parts surface cleaning, plating, rinsing, stripping, and waste treatment. Their pollution prevention program also incorporates: consideration of waste reduction potential when designing new equipment and facilities, active participation of employees in carrying out the program's goals, and strong leadership by Stanadyne personnel in promoting these activities among other industries.

Parts Surface Cleaning

The cleaning of brass and copper parts which have been brazed together is accomplished in a bright dipping operation. This operation -- cleaning metal surfaces with acids and oxidizers -- is a major waste producer. A sulfuric/peroxide bright dip is used to avoid the high sludge production associated with chromic acid bright dip. A side tank, which will be used as an electrolytic cell, has been ordered for plating out the copper which builds up in the bath. Redesign of the racks which carry the parts will also be considered in 1985.

Plating Operation

Plating is accomplished on two return-type machines (Udylite Cyclemasters). As parts are lifted out of the plating bath, the solution that clings to the part is called dragout, which becomes pollution in the rinsing process. Minimization of dragout should be involved in the investigation of four areas: solution makeup, machine motion, part and rack design, and solution maintenance.

Solution Makeup

Differently shaped parts require plating baths of different chemical makeups to achieve required thickness and deposit qualities. In formulating their cyanide copper- and chrome-plating baths, Stanadyne determined that lower concentrations of chemicals could be used. By running the potassium cyanide concentration at 2.5 ounces per gallon, instead of 3.5 ounces per gallon, the cyanide dragout concentration has been reduced by 28 percent, without any adverse affect on plating quality. Savings accrue from using less cyanide and less bleach for cyanide destruction. In the chrome baths, the chromic acid levels are maintained at about 29 ounces per gallon instead of 32 ounces per gallon. This 9-percent reduction results in savings due to reduced raw material requirements. These materials include chromic acid, sodium bisulfite (used to reduce hexavalent chrome to trivalent chrome), and sodium hydroxide (used to precipitate metals). Sludge handling costs are also reduced.

Machine Motion

The motion of the Udylite Cyclemaster plating machine is repetitious and automatic. Work is moved through a tank, lifted out, indexed over, and lowered into the next tank. By introducing a pause before indexing in the up position, a significant portion of the dragout is able to drip back into the process tank. Specific savings due to this change have not been calculated. In making this change, care was taken to avoid an excessive increase in "hang time". Depending on the base metal and on the process, dry-ons or passivation of nickel could result.

Part and Rack Design

Stanadyne adopted a program of rack redesign for minimization of dragout, after attending the "Pollution Prevention Pays in the Electroplating Industry" workshop in 1983 (sponsored by the state of North Carolina). Although savings are difficult to calculate, significant savings in process chemicals, pollution control chemicals, dropped parts, and in future rack-coating costs have been achieved. The first consideration for rack redesign was a change from horizontal to vertical orientation. Instead of running tips off the horizontal cross-members, tips were connected directly to vertical splines. Several advantages were gained by this change:

  1) dragout was reduced because the solution slides off the rack, instead of clinging to horizontal cross members;

  2) rack area was reduced, providing less surface area for dragout to cling to;

  3) fewer electrical resistance points, since splines are direcly connected to rack hooks;

121

4) reduced shading of low current density areas by each rack spline;

5) reduced plating on the backs of parts which are not exposed surfaces, therefore requiring less protection for these parts;

6) rack recoating costs were reduced because there was less rack area.

The changes in rack design were undertaken during the regular rack-recoating process. Initial cost was somewhat higher than a simple recoat, but the change will quickly pay for itself.

Wide shelf areas, cups, and blind holes were eliminated wherever possible. The orientation of each part on the rack was reviewed. By angling parts to one side, "cups" were eliminated, and dragout was reduced. The longest lines of each part are racked vertically, whenever possible, to allow dragout to slide off more easily.

Solution Maintenance

Stanadyne reviewed the plating solution filtration process for waste reduction opportunities. Plating solutions must be filtered to produce high-quality finishes which are free from roughness. Filter media must be changed whenever pressure builds up and retards the rate of filtration. During filter changes, spillage or outright dumping can become a significant source of pollution and/or process water treatment problems. In late 1982 and early 1983, all sleeve-type filters were replaced with horizontal plate filters, which use paper disks as filter media. The new Mefiag filters are manufactured by Baker Brothers. The payback period for the new filters was less than two years. The most startling savings occurred in the cost of media for filter changes. The reduction in the cost of media per filter change resulted in a 93 percent direct cost savings. Solution losses, as well as the labor time needed to change the filters, were also reduced.

A method for reducing solution losses during filter changes was recommended by Dave Gibson of Baker Brothers; it was instituted at Stanadyne in 1984. An air hose from the low-pressure air blower is connected to the filter's bleed-off hose. Air is introduced into the filter, and the solution can be blown into a slurry or storage tank. The longer the filter is blown down, the drier it becomes. Using this method, solution losses were nearly eliminated, and filter changes are easier and safer. This is especially helpful when handling cyanide solution.

Stanadyne has observed a 20-percent reduction in chemical costs per hour as a result of the zinc die-cast plating modifications. This translates into over $10,000 in cost savings per year.

Dragout Recovery and Rinsing

By capturing dragout in still tanks before rinsing and recycling it to the plating tanks, valuable chemicals can be saved, and treatment and disposal costs eliminated. Dragout recovery requires strict control of the contaminants in the plating solution to prevent concentrating them, which would result in plating defects.

A simple dragout recovery system was installed on the nickel plating machine for zinc die-castings. Less than $1,000 was invested for a storage tank. The dragout is recycled once a week. The nickel is recovered at a concentration of 3.2 ounces per gallon, which results in a 30-pound-per-week savings in raw materials. At the current market price of $2.80 per pound, this system saves the firm $4,200 worth of nickel per year. The recovery operation also reduces the generation of nickel sludge by 9,500 pounds per year.

Plans are now underway to recycle chrome dragout. The system will not recover 100 percent of the lost chrome, but will attempt to reach an 80- to 90-percent recovery rate. Since the evaporation from the chrome tank is too low to allow the direct return of chrome dragout, an evaporator tank will be used to reduce the dragout volume before pumping it directly back to the chrome tank. Sulfate and cation concentrations will be monitored for possible problematic buildups.

Counterflow rinsing and flow controls to rinse tanks are standard practices at Stanadyne. Water reuse, both before and after treatment of the tanks, is being considered, but has not yet been instituted.

Waste Treatment

The reduction of sludge volume is achieved by reducing the water content of the sludge. Sludge results from the removal of metals from rinse waters; the sludge is in the form of metal hydroxide, a hazardous waste. At Stanadyne, sludge volume reduction is achieved through the use of a filter press. Dewatering with the press results in a 25- to 30-percent-solid sludge. The reduced sludge volume decreases shipping and disposal costs.

A further reduction is currently being pursued through the purchase of a JWI (J-Mate) sludge dryer. This piece of equipment, which is currently on order, represents new technology introduced in July 1984. The dryer will drive the remaining 65- to 70-percent of water of hydration from the sludge. The residual solids will have a significant metal content, 30 to 40 percent, which may be recoverable, if not now, then in the future.

Wastewater Treatment Improvements

Adding a sand filter to their wastewater treatment system has helped Stanadyne produce a higher quality wastewater. The DynaSand filter, which cost about $40,000, incorporates the new idea of a moving, continuously cleaned sand filtration bed, which significantly reduces the volume of water required for backwashing. Other small changes were made in the system to improve water quality: an improved polyelectrolyte pumping system, a changed procedure for handling spent cleaner, and installation of metering pumps on collection pits to eliminate system overloads.

In 1983-84, Stanadyne increased its plating on zinc by 15 percent, its plating on brass by 56 percent, and its bright dipping by 35 percent. In the face of these increases, the company's pollution prevention modifications still managed to reduce wastewater treatment chemical costs by 20 percent, and allowed only a 2-percent increase in sludge generation. The chemical cost reductions resulted in a net savings of $5,000 per year.

Stripping

Stripping nickel and chrome off brass parts is a rework operation designed to save the expensive brass base metal. Until recently, immersion stripping was used to remove nickel and chrome. An immersion stripper loses efficiency as it is used. The bath slows down as metal loading increases, and it ultimately must be "dumped" and remade.

A new electrolytic nickel stripper (Udystrip 8000 by Udylite) has been installed which deplates nickel from the brass parts. Nickel sludge builds up on the electrodes and on the tank bottom; it can be removed by pulling the electrodes, or by decanting the bath, removing the sludge, and pumping the bath back to the tank. A volume of liquid, equal to the volume of sludge removed, is then added to the bath. This bath does not have to be dumped, and rinse waters are waste-treatable. The new stripper resulted in an 81-percent reduction in chemical costs per hour, which translated into a savings of approximately $5,000 per year. Having less waste to treat and dispose of, and having a more scientific and predictable operation also led to cost savings. Initial capital costs were higher, since a rectifier had to be purchased, but long-term costs were significantly reduced, with savings greater than 50 percent.

Chrome is stripped in a 75-percent muriatic acid dip, immersion stripper.

Housekeeping

Recognizing that plating a defective part creates the same amount of waste as plating a good one, Stanadyne began preinspecting work to be plated. Parts that would obviously be scrapped are eliminated from the plating operation, resulting in direct waste reduction. Elimination of leaks also resulted in waste reduction.

Plant Expansion

During 1983, Stanadyne expanded its plant, increasing its capacity by 25 percent. Additional plating and degreasing capacity was included in the expansion. Each of these operations is a source of hazardous waste generation; extensive planning of equipment layout, operation, and design was undertaken to maximize pollution prevention.

Plating Machine

In planning the layout for a new plating machine, an effort was made to minimize pollution sources and rinse water flow, and to include features that would ease future recovery opportunities. The plating machine was designed with double counterflow rinses and, where possible, triple counterflow rinses. In addition, the plating tanks (nickel and chrome) are followed by dragout tanks to allow some direct recycling of solutions which would otherwise be wasted. The design of the floor layout included additional space in a chrome reduction unit pit; this will permit the installation and cost-effective operation of a simple chrome recovery system currently under design.

Degreaser

A planning approach similar to the one used for the new plating machine was used during the purchase of a new totally-enclosed, automatic degreaser. After analyzing the different machines that were available, a unit manufactured by Finishing Equipment, Inc., of St. Paul, Minnesota, was selected. Pollution prevention and energy efficiency were major factors in

the decision to purchase this particular unit.   Stanadyne
selected the unit based on the following features:

1) The chiller system is run continuously to keep the
solvent in the machine, where it belongs. Major solvent losses
usually occur when a degreaser is not in operation. By running
the chiller system continuously, these losses are eliminated.

2) No water is used in the system for cooling or rinsing.
This reduces water costs, and ultimately helps the community, by
limiting the need for expansion of the municipal wastewater
treatment plant.

3) Waste heat from the chiller is used to heat the
degreaser unit, resulting in energy savings and conservation of
resources.

4) The machine design offers two major pollution prevention
features: (a) the parts are continuously rotated, which
facilitates drainage and prevents solvent dragout; (b) the chip
removal system eliminates solvent loss during chip removal, and
eliminates a major safety hazard for employees.

5) Using methylene chloride in the degreaser system reduces
solvent waste because the chemical stability allows it to be
distilled many times before it degrades. It also has a low
boiling point, which saves energy.

6) The system is designed with a large distillation
capacity. This allows the still bottoms to be reduced to a 6- to
10-percent solvent content.

The features which led to the purchase of this unit are
paying dividends. Solvent usage has dropped from an average of 7
drums per month to only 1.5 drums per month. Air emissions and
hazardous waste generation have been reduced substantially.

Management Initiatives

Stanadyne believes that one of the major strengths of their
pollution prevention program is the involvement of all the
employees who work in plating and pollution control operations.
The staff is trained to appreciate the contribution that each job
makes to the overall success of the program.

The enthusiastic backing of pollution prevention projects by
the firm's corporate management has led to company participation
in numerous environmental programs, boards, and committees to
explain and promote Stanadyne's activities.   Stanadyne was
instrumental in organizing workshops throughout North Carolina on
"Making Pollution Prevention Pay in the Electroplating and Metal
Finishing Industries".   They have also opened their plating

126

operation to professional associations for plant tours. In 1983, a film crew, from the University of North Carolina television station , visited Stanadyne to film a report on hazardous waste management in North Carolina. Plant manager John Carney led a tour of the plant and answered questions for the program.

In September of 1982, George McRae, plating manager at Stanadyne, proposed the formation of a Lee County Industrial Waste Committee, now called the Lee County Waste Managers Association. The purpose of the association is to trade ideas and information, coordinate waste shipments, and to offer an pportunity for discussion of issues related to the Lee County Waste Board. Initially, the Association met monthly to give the waste managers a chance to become acquainted, and to discuss ideas and problems. A tangible benefit resulting from the Association was the initiation of shared waste shipments. During the past year, Stanadyne has shared waste loads with another Lee County company on four occasions. This arrangement has allowed both companies to move their waste in a cost- effective manner, in less than 90 days. Stanadyne and the company with which it shares loads have both been delisted as storers under the RCRA guidelines. Savings occur because transportation of waste is computed on a per-mile-hauled basis. This means that the charge for transportation is the same whether the truck carries 20 drums or 80 drums. At $3.20 per loaded mile, it definitely pays to send a full truckload. George McRae currently serves as association chairman.

PROCESS DIMENSIONS

WASTES GENERATED	Reduced by 46 percent (plating bath compositions, filter media change, dragout recovery system, sludge filter press, stripping operation, cleaning)
PRODUCTIVITY	Improved (rack design)
PRODUCT QUALITY	Improved (rack design)
DOWN TIME	Reduced (filter media change, stripping operation)
FACILITIES	Expanded, cost information not provided
EQUIPMENT	New plating solution filters, sludge filter press, nickel stripping system, dragout recovery tank, process water filtration system. Cost information not provided.
RAW MATERIALS	$15,000 saved per year (cyanide, bleach, chromic acid, sodium bisulfite, sodium hydroxide, other plating solution chemicals, stripping chemicals)
WATER	Costs minimized (new degreaser); complete information not provided.
ENERGY	Information not provided

WASTE DISPOSAL	Reduced transport and disposal costs
POLLUTION CONTROL	$5,000 saved per year
PERSONNEL/MAINTENANCE	Reduced rack coating costs, reduced filter media replacement costs, initial rack coating costs increased during redesign
NET BENEFITS	Annual cost savings, improved product quality, reduced regulatory pressure

Case Study 25

TYPE OF INDUSTRY: Fabrication of Pipe Fittings

NAME OF FIRM: Elkhart Products Division, Inc.
                   (Lancy International, consulting firm, William
                   McLay 412/452-9360)

LOCATION: Elkhart, Indiana

SIC: 3471

```
*┌─────────────────────────────────────────────────────────┐*
*│                                                           │*
*│                   MODIFICATIONS                           │*
*│                                                           │*
*│   1) REDESIGN OF BRITE PICKLING PROCESS TO REDUCE WASTES  │*
*│                                                           │*
*│   2) ELECTROLYTIC RECOVERY OF COPPER                      │*
*│                                                           │*
*└─────────────────────────────────────────────────────────┘*
```

Abstract

    Elkhart Products, Inc., manufactures pipe fittings. Under
the guidance of Lancy International, they have modified their
bright-dipping and passivating-line wastewater treatment system.
It was replaced with a newly developed pickling process that
utilizes sulfuric acid and a proprietary, stabilized hydrogen
peroxide. By modifying the equipment operation, Elkhart was able
to remove a cyanide dip and chromic acid dip from the process
line. They also recover metal, using an electrolytic recovery
cell, currently recovering copper at the rate of 1,000 to 1,500
pounds per month.

ANNUAL COST SAVINGS: $120,000

PAYBACK PERIOD: 0.5 years

TIME TO IMPLEMENT: 6 months

TRANSFERABILITY: Widely applicable in the metalworking
                        industry; however, proprietary
                        elements of the process may limit its
                        reproduction. The process efficiency
                        of electrolytic recovery with other

metals and wastewater compositions is variable and must be considered.

ENVIRONMENTAL AND
HEALTH BENEFITS: Eliminated risks of surface- and groundwater contamination by hazardous wastes; conservation of water and metal resources; reduced chances of worker or non-worker contact with hazardous waste.

Project Description

    Elkhart Products, Inc., fabricates pipe fittings. Lancy International was contracted to design and install a treatment system that would meet a number of criteria. Elkhart required a system that would comply with both present regulations and more stringent future regulations; that would not lower product quality; that would have the lowest total waste treatment costs; and that would recover the large amounts of copper that were being discharged down the drain. Elkhart's original system consisted of a conventional alkaline cleaner and rinse, followed by a cyanide dip to remove some of the oxides, a water rinse, a chromic acid bright dip/passivation line, three cold water rinses, and a hot water rinse.

Lancy responded with a total redesign of the entire process system; the new design emphasized the reduction of wastes requiring treatment. Lancy replaced the existing cleaning sequence with a totally different process that utilizes sulfuric acid and a proprietary stabilized hydrogen peroxide. The line was converted to include an alkaline cleaner, two water rinses, the new sulfuric acid hydrogen peroxide bright pickle dip, and a countercurrent rinse. The final rinse water is regenerated and reused by precipitating the copper as cuprous oxide and/or copper hydroxide. The bright pickle dip is also regenerated by transferring the nearly saturated contents of the pickle tank to a larger reservoir, where the copper sulfate is precipitated and removed by a cyclonic separator. The pickle dip efficiency is continually maintained by the addition of sulfuric acid and the proprietary stabilizers. The copper sulfate crystals, the cuprous oxide, and the copper hydroxide "wastes" are introduced into an electrolytic cell, redissolved, and plated out as No. 1 copper scrap.

    By implementing these modifications, Elkhart was able to remove the cyanide and the chromic acid bright pickle dips from the process line. They were also able to regenerate their new bright pickle dip, and to remove and recover the copper at a rate of 1,000 to 1,500 pounds per month.

PROCESS DIMENSIONS

WASTES GENERATED	Reduced by 182,000 gallons per year
PRODUCTIVITY	No change
PRODUCT QUALITY	No change
DOWN TIME	No change
FACILITIES	$30,000 one-time cost
EQUIPMENT	$30,000 one-time cost
RAW MATERIALS	$1,770 saved per year
WATER	Information not provided
ENERGY	Information not provided
WASTE DISPOSAL	$18,200 saved per year
POLLUTION CONTROL	Reduced
PERSONNEL/MAINTENANCE	Information not provided
NET BENEFITS	Resource conservation, cost reduction

For more information see "Company Approaches Zero Pollution Discharge," by B. Mottweiler and P.L. Veit, in Plating and Surface Finishing, December 1979.

Case Study 26

TYPE OF INDUSTRY: Electroplating Job Shop

NAME OF FIRM: Pioneer Metal Finishing, Inc.
                 (Harry DeSoi 609/694-0400)

LOCATION: Franklinville, New Jersey

SIC: 3471

```
*  ───────────────────────────────────────────────────────  *
*                                                             *
*                       MODIFICATION                          *
*                                                             *
*     REPLACEMENT OF A SINGLE-PASS, CONTINUOUS WASTE TREATMENT *
*        SYSTEM WITH A CLOSED-LOOP, BATCH TREATMENT SYSTEM     *
*                                                             *
*  ───────────────────────────────────────────────────────  *
```

Abstract

Pioneer was using a single-pass system to treat their
plating wastewaters. Because the system had no large end tank to
correct minor upsets in the system, concentrations of metals and
other inorganic chemicals in the effluent regularly exceeded
water quality regulations. The effluents were killing many
species of plants and animals in the receiving marsh. Pioneer
solved this problem by replacing the old single-pass, process
water treatment system with a closed-loop, batch process system
manufactured by Zerpol. As a result of the new equipment design,
the company now uses only 500 gallons of water per day (makeup
due to evaporation), in contrast with the previous 12,000 gallons
per day, and has experienced a 20-percent reduction in the amount
of sludge produced. Plans are being made to sell the sludge to a
company for use as a filler in the manufacture of concrete
products.

ANNUAL COST SAVINGS: $52,460

PAYBACK PERIOD: 3 years

TIME TO IMPLEMENT: 1 year

TRANSFERABILITY: Widely applicable to metalplating
                    operations

133

ENVIRONMENTAL AND
HEALTH BENEFITS: Reduced threat to marsh ecosystem and
                                to adjacent surface and groundwater;
                                conservation of metal resources;
                                improved health and safety conditions
                                for plant personnel.

Project Description

Pioneer Metal Finishing, Inc., is an electroplating operation. Rinse waters contaminated with metals and cyanide pass into tanks fitted with a continuous monitoring and treatment system. The theory was that chemicals would be added in response to signals from various probes and sensors, in order to oxidize, reduce chromates, and precipitate metals. After centrifugation to remove the solids, the clean water was released into a nearby marsh. Unfortunately, there were occasional minor upsets in the system; while the operator was determining what had happened, the concentrations of certain pollutants, in the effluent leaving the plant, would rise too high, violating state water quality regulations.

In order to meet stringent new regulations, Pioneer installed a closed-loop, batch treatment system for their rinsewater. The wastewater is now captured in large treatment tanks, allowing more time for the reactions to proceed to completion. This has resulted in less chemical use and less sludge production. After the solids are removed, the water is reused. There have been no problems with excessive salt build-up. Recycled water is not used for rinses that precede plating with nickel, trivalent chromium, nor is it used in the hot water rinse after the chromium dip. These operations require high-quality distilled water.

Chemicals and water were also saved by:

1) Educating employees about opportunities to reduce waste;

2) Educating customers to locate the drain hole, on their products to be plated, so that dragout is minimized;

3) Properly designing the plating rack to limit dragout.

The savings involved in these process modifications are impressive:

   o   Chemical treatment costs were reduced by 50 to 75 percent
   o   Sludge production decreased by 50 percent
   o   Water bills were reduced by 90 percent

134

PROCESS DIMENSIONS

WASTES GENERATED	Reduced by 40,000 gallons of sludge per year
PRODUCTIVITY	No change
PRODUCT QUALITY	No change
DOWN TIME	Reduced
FACILITIES	$60,000 one-time cost
EQUIPMENT	$150,000 one-time cost
RAW MATERIALS	No change
WATER	11,500 gallons saved per day
ENERGY	$100 saved per year
WASTE DISPOSAL	$58,460 saved per year
POLLUTION CONTROL	$29,400 saved per year
PERSONNEL/MAINTENANCE	$10,200 saved per year
NET BENEFITS	Minimization of waste management cost

Case Study 27

TYPE OF INDUSTRY: Farm and Construction Equipment Manufacture

NAME OF FIRM: Deere and Company
                      (Mike McGuire 309/752-5435)

LOCATION: Moline, Illinois

SIC   3520 & 3530

```
*  ------------------------------------------------------------  *
*                         MODIFICATIONS                          *
*                                                                *
*  1)   DEVELOPMENT OF A HAZARDOUS WASTE TASK FORCE              *
*                                                                *
*  2)   COMPREHENSIVE WASTE TREATMENT FACILITY TO RECLAIM AND   *
*       DETOXIFY SELECT HAZARDOUS WASTES                         *
*  ------------------------------------------------------------  *
```

Abstract

Deere and Company reduced its hazardous waste by 80 percent in volume and by over 99 percent in level of risk through two key actions:

1) Corporate implementation of a comprehensive hazardous waste management initiative at each manufacturing unit;

2) Construction of an on-site facility to treat hazardous liquid waste; the facility is able to reclaim waste organic oil compounds and metals from process water, and produces a non-hazardous sludge. The reclaimed oil is sold to an oil recycling firm or is reused for machining processes.

ANNUAL COST SAVINGS: $155,750

PAYBACK PERIOD: 2.5 years

TIME TO IMPLEMENT: 1.5 years

TRANSFERABILITY: Applicable to any firm that must
                      manage large volumes of wastewater
                      containing oils or metals.

```
ENVIRONMENTAL AND
HEALTH BENEFITS:                  Reduced threat to air and groundwater
                                  quality; conservation of metal,
                                  energy, and water resources; improved
                                  health and safety conditions for plant
                                  personnel.
```

Project Description

Deere and Company manufactures farm and construction equipment. In 1980, the company formed a Hazardous Waste Task Force comprised of employees from all levels, and charged it with four goals: (Because of proprietary rights on the technology, only a general description follows.)

1) To develop options for establishing a comprehensive strategy for handling hazardous waste materials at each manufacturing unit;

2) To define the degree of hazardous waste treatment that should be conducted and which units should be involved;

3) To define programs that will assure regulatory compliance;

4) To eliminate the need for hazardous waste disposal whenever possible.

After extensive study, the Hazardous Waste Task Force concluded, and top management concurred, that most hazardous waste disposal activities could be eliminated through the following activities:

1) Revision of manufacturing procedures to minimize hazardous waste generation;

2) Research and development of methods to detoxify hazardous waste;

3) Construction and interfacility use of a single, high-technology facility for liquid waste treatment, that would produce a nonhazardous sludge and meet effluent standards;

4) Use of vendors for solvent reclamation activities;

5) Continued audit of each unit's hazardous waste management activities to assure efficient progress in reducing and eliminating toxic materials.

Since 1980, Deere has reduced its hazardous waste disposal by 80 percent; they anticipate a 98-percent reduction in hazardous wastes in 1985. The company continues to closely monitor all of the systems to ensure that they are making timely progress toward their goal of maximum reduction of hazardous waste.

The central aspect of Deere's comprehensive strategy was the development of the liquid waste treatment facility. It has the following capabilities:

1) Oil reclamation from 15- to 60-percent oil/water mixtures;

2) Treatment of concentrated heavy metal wastes;

3) Production of sludge that does not test toxic under the RCRA extraction procedure.

The facility can treat over 2 million gallons per year of the following types of hazardous wastes: electroplating sludges, spent electroplating baths, painting wastes, caustic paint-stripping wastes, kolene sludge from casting cleaning, and miscellaneous acid and alkaline wastes. Also, 1,380,000 gallons per year of waste oils and oil/water mixtures can be treated such that oil is reclaimed for sale to an oil recycling firm, or is reused for machining processes. Deere and Company currently has a patent pending on the process and is willing to license its use to other industries.

PROCESS DIMENSIONS

==

WASTES GENERATED	Reduced hazardous wastes by 330,000 gallons per year
PRODUCTIVITY	No change reported
PRODUCT QUALITY	No change reported
DOWN TIME	No change reported
FACILITIES	$1,900,000 one-time cost
EQUIPMENT	Included under "facilities" parameter
RAW MATERIALS	Savings due to reuse of oil, dollar figure not provided
WATER	Information not provided
ENERGY	Information not provided
WASTE DISPOSAL	$230,000 saved per year
POLLUTION CONTROL	$74,250 cost per year
PERSONNEL/MAINTENANCE	No change
NET BENEFITS	Minimization of waste management cost

==

For more detailed information, refer to "Corporate Strategy and Actions for Reducing Hazardous Waste Disposal Requirements," by R.D. Grotelueschen, Manager of Environmental Control, Deere and Company.   Published in Pollution to Profit: Reducing Industrial Wastes in Illinois, proceedings from The Illinois Conference on Industrial Waste Reduction, Document No. HW84/02, June 1984, pp. 8-18.

Case Study 28

TYPE OF INDUSTRY: Manufacture and Metal Finishing of Stationary
                   Power Tools

NAME OF FIRM: Emerson Electric Company
                   Special Products Division
                   (Layton Schuh 704/837-5101)

LOCATION   Murphy, North Carolina 28209

SIC: 35, 36

```
*************************************************************************
*                                                                       *
*                          MODIFICATIONS                                *
*                                                                       *
*    1)  REPLACEMENT OF CONVENTIONAL, ORGANIC-SOLVENT-BASED             *
*        ELECTROSTATIC PAINT SYSTEM WITH A WATERBORNE                   *
*        ELECTROSTATIC PAINT SYSTEM                                     *
*                                                                       *
*    2)  INSTALLATION OF A COMPUTERIZED ROBOTIC ELECTRO-                *
*        PLATING SYSTEM                                                 *
*                                                                       *
*    3)  INSTALLATION OF AN ULTRAFILTRATION SYSTEM FOR                  *
*        OIL AND PROCESS CHEMICAL RECOVERY                              *
*                                                                       *
*    4)  INSTALLATION OF A COMPUTERIZED ENERGY MANAGEMENT               *
*        SYSTEM                                                         *
*                                                                       *
*    5)  DEVELOPMENT OF A COMPREHENSIVE PROGRAM FOR CHEMICAL            *
*        WASTE MANAGEMENT                                               *
*                                                                       *
*    6)  INTRODUCTION OF AN INCENTIVE PROGRAM FOR NEW COST-             *
*        REDUCTION OR PRODUCT IDEAS                                     *
*                                                                       *
*************************************************************************
```

Abstract

     Emerson Electric Company has made a series of modifications
which have significantly decreased operating costs; increased
productivity; reduced solvent, paint solids, and plating chemical
costs; conserved energy; improved product quality; provided a
safer working environment; and conserved raw materials. Several
of the improvements involved the purchase of new equipment which
allowed the automation of their electroplating process and their
energy management system. Conversion from an organic-solvent-

based paint system to a water-based system reduced paint solvent and paint sludge losses by 40 percent. The addition of an ultrafiltration system for their wastewater stream has allowed the recovery of oil and process chemicals. As a result, the load to the wastewater treatment system has been reduced, and $11,000 in raw materials are returned to the production processes. A management initiative to develop a program to oversee all waste management activities has also been instituted. Several housekeeping improvements have resulted from this program; they have yielded significant cost and environmental benefits.

ANNUAL COST SAVINGS: $1.8 million

PAYBACK PERIOD: Modification #1:   1.1 years
                         Modificaiton #2:   1.1 years
                         Modification #3:   2 years
                         Modification #4:   5 years

TIME TO IMPLEMENT: 1 year

TRANSFERABILITY: The most important features of these
                         modifications are the comprehensive
                         nature of Emerson's pollution
                         prevention activities, and the
                         significant savings they were able to
                         realize as a result of their efforts.
                         Changes made in several finishing
                         processes reduced the amount of wastes
                         requiring treatment in the company's
                         on-site system. Several of the modifi-
                         cations required significant capital
                         expenditures; these might not be cost-
                         effective at smaller facilities. The
                         housekeeping activities demonstrate the
                         savings that can be achieved by consis-
                         tently monitoring all phases of a manu-
                         facturing operation, and by keeping
                         abreast of new product development.

ENVIRONMENTAL AND
HEALTH BENEFITS: Conservation of material resources;
                         conservation of energy; elimination of
                         oil spill risks; reduction of oil and
                         chemical loading to the effluent
                         stream; reduced employee exposure to
                         organic solvents, acids, and caustics.

Project Description

Emerson Electric Company manufactures stationary power tools; these require both metal finishing and painting operations. Chemical processes in operation include:

1) An electrostatic paint system, with a seven-stage-metal pretreatment line, that finishes 75,000 square feet of parts per day;

2) A black penetrate and zinc electroplating line that finishes 10,000 pounds of small parts per day;

3) An aluminium die-casting facility that produces 30,000 pounds of parts per day;

4) A paint stripping line that strips paint from 1,000 square feet of parts per day;

5) A two-stage, small-parts washer that cleans 3,000 pounds of ware per day;

6) A vibratory finishing line that processes 4,000 pounds of parts per day;

7) A machining center that contains 2,000 gallons of machine coolant, and generates 10,000 pounds of miscellaneous sludge and 75 gallons of spent chemicals per day.

When Emerson installed an automated, metal electroplating system to replace their manual system, annual productivity increased by $200,000, and system down-time decreased from 8 to 4 percent. Chemical consumption has decreased by 25 percent, resulting in an annual reduction of $8,000 in raw material costs. Water costs have been reduced by $1,100 per year, and plating wastes, including acids, caustics, and oils, have decreased from 450 to 360 pounds per day. Treatment costs for the process water water used in the plating operation have been reduced by 25 percent. Annual savings of $35,000 in personnel and maintenance costs are attributable to the new system. The automated system has also eliminated worker exposure to acids and caustics; exposure was previously unavoidable with the manual operation.

A second major revision to the metal-finishing operation was the substitution of an anodic electrostatic immersion paint system that is water-based for an existing organic-solvent-based paint system. As a result, the quality of the paint application has been greatly improved, and annual productivity has increased by $1,000,000. Down time for the new system has decreased from 3 to 1 percent. The water-based system allows 99.5-percent recovery and reuse of paint. Consequently, raw material costs for paint have decreased by $600,000 per year. Instead of

143

generating 3,000 pounds per day of aromatic, chlorinated waste solvent and 70 pounds per day of paint solids, the company now generates only 150 pounds of non-reactive solvent per day, and only two pounds of paint solids per day. As a result of these changes, hazardous waste disposal costs have been reduced from $10,000 per year to only $300 per year. With the new paint system, personnel and maintenance costs have been decreased by 40 percent, and worker exposure to organic paint solvents has been eliminated.

The wastewater from the manufacturing process, which contains oil and process chemicals, is treated on-site in a 55,000 gallons per day, NPDES permitted wastewater treatment plant. The plants processes include neutralization, polymerization, dissolved air flotation, sand filtration, and carbon adsorption. The effluent is discharged to a Class II stream. The addition of a $65,000 oil skimmer and ultrafiltration system has allowed the recovery of oil worth $8,000 per year. Twenty-five hundred gallons of reusable alkaline cleaning solution are also recovered each month, saving approximately $3,000 per year. About 370 pounds of BOD are removed each month by the ultrafiltration unit. The reduced organic loading to the treatment system has cut water treatment costs by $10,000 per year. The new filtration process has also saved the firm from having to expand their water treatment capacity.

In addition to these specific process modifications, Emerson has developed a broad-based program for chemical waste management. The program was initiated throughout the Special Products Division of Emerson Electric, with the president and vice president of operations taking personal action. The program was initiated because the company had a number of specific waste management problems, including:

1) Continual oil spills at the scrap loading site. The oil contaminated the surrounding surfacewater; this water ultimately drained into the adjacent creek. Approximately 20 gallons of water-soluble, petroleum-based oil were released to the environment each week.

2) Continual oil spills from the aluminum die-casting facility contaminated an adjacent waterway. Approximately 45 gallons of water-soluble and water-insoluble petroleum-based oils and synthetic oils were released to the environment each week.

3) Weekly episodes of unauthorized dumping of process chemicals into the water treatment system. This practice overloaded the treatment system's organic removal capacity, and subsequently resulted in inefficient metal removal.

4) Monthly overloads of the wastewater treatment system caused by the dumping of a concentrated solution of spent alkaline cleaner. As a result of this practice, the effluent exceeded the allowable limits for organics and metals.

5) Improper labeling and storage of hazardous and non-hazardous chemical wastes; if spilled these wastes could enter an adjacent waterway. At one time 20,000 pounds of unidentified chemical wastes were being stored outside.

6) Excessive use of chlorinated solvents in degreasing operations; these solvents affected the wastewater treatment system. Approximately 1,800 pounds of solvent were used per month; half of this was introduced to the wastewater treatment plant.

7) Improper use of zinc chromates as an algacide in the evaporative cooling water systems. Each minute approximately four gallons of continual blowdown, containing zinc chromates, were discharged to the storm drain.

After beginning their waste management program, the Company reports that:

1) They invested $25,000 in a die-cast sump and pumping facility and in a scrap-loading facility. Previously all oils from die-casting and at the scrap-loading site were lost; approximately 65 gallons per week are now recovered. The improvements save approximately $8,000 per year.

2) They discontinued all unauthorized dumping of waste chemicals to the wastewater treatment plant. Operation of the treatment facility has improved, and effluent guidelines are being protected.

3) They invested $3,000 in drum storage and handling. All hazardous chemical wastes are properly labeled, stored, manifested, and managed by an EPA-approved vendor. No hazardous wastes are stored outside; they are properly stored, and are specially treated or picked up by appropriate vendors.

4) They reduced degreasing with chlorinated solvents by 80 percent. A steam cleaner was purchased for $5,000, and degreasing is accomplished by using an alkaline detergent and steam. This has reduced the toxic organic loading to the wastewater treatment plant, and the treated effluent is meeting the currently proposed guidelines.

5)   They reformulated the algacides in the cooling water systems and removed the zinc chromates.   Chlorinators for the cooling towers were purchased for $1,000.

The management program is ongoing and is monitored by the in-plant process engineer. The continuing goals of the program are:

1)   To improve the daily operations and effluent of the wastewater treatment plant by eliminating unauthorized dumping of spent chemicals;

2)   To secure the safety of employees by educating them about the proper disposition of all waste process chemicals;

3)   To prevent chemical spills on the plant's grounds;

4)   To comply with the pertinent RCRA regulations, and with other relevant environmental laws.

Process technicians maintain daily log sheets. All plant personnel with designated responsibilities in the program, are properly informed, and have access to the management booklet. Any violations of the stated procedures are reported to the plant manager.

Emerson also maintains an incentive awards program to encourage new ideas for cost savings or for development of new products. Items such as color televisions, radios, and stereos, are awarded to acknowledge employee contributions.

A final modification reported by Emerson was the installation of a $100,000 computerized energy management system; this saves the firm $145,000 in electrical costs per year, and has reduced personnel and maintenance costs by an additional $2,000 per year.

PROCESS DIMENSIONS
==

WASTES GENERATED	1) Eliminated 3,000 pounds of aromatic chlorinated solvent waste per day (paint system) 2) Reduced waste paint solids, from 70 to 2 pounds per day (paint system) 3) Reduced plating acid, caustic, and oil from 450 to 360 pounds per day (plating system) 4) Eliminated 65 gallons of waste oil per day (ultrafiltration) 5) Reduced chlorinated waste solvents from 900 to 180 pounds per month (degreasing) 6) Eliminated zinc chromate discharge to storm drains (cooling water)
PRODUCTIVITY	1) Increased $1,000,000 per year (paint system) 2) Increased $200,000 per year (plating system)
PRODUCT QUALITY	Improved paint coverage, uniformity of coating thickness
DOWN TIME	1) Reduced from 3 to 1 percent by new paint system 2) Reduced from 8 to 4 percent by new plating system

FACILITIES	$200,000 one-time cost
EQUIPMENT	$674,000 one-time cost
RAW MATERIALS	$642,000 saved per year
WATER	$2,200 saved per year
ENERGY	$145,000 saved per year
WASTE DISPOSAL	$ 52,700 saved per year
POLLUTION CONTROL	$380,000 cost per year
PERSONNEL/MAINTENANCE	$157,000 saved per year
NET BENEFITS	Annual cost savings, improved on-site waste treatment efficiency, reduced regulatory pressures, decreased worker exposure to organic solvents

Case Study 29

TYPE OF INDUSTRY: Small Appliance Manufacture

NAME OF FIRM: Hamilton Beach Division of Scovill, Inc.
(Guy Tilford 919/592-0121)

LOCATION: Clinton, North Carolina

SIC: 36

```
*──────────────────────────────────────────────────*
*                 MODIFICATIONS                     *
*                                                   *
*   1)  SOLVENT RECYCLING                           *
*                                                   *
*   2)  SUBSTITUTION OF WATER-BASED CLEANERS FOR ORGANIC-  *
*       SOLVENT-BASED CLEANERS                      *
*──────────────────────────────────────────────────*
```

Abstract

The Hamilton Beach Division of Scovill, Inc., manufactures small electric appliances. Their operation requires 1,1,1-trichloroethane solvent to degrease metal stampings. Solvent wastes are collected in 55-gallon drums; the drums are housed in a storage building designed to contain possible spills. The drums are picked up by a company that distills 1,1,1-trichloroethane from the wastes. This company returns the solvent to Scovill for reuse. The recycled solvent costs about half as much as the virgin material. Scovill has also experimented with water-soluble synthetic cleaners as a material substitution for organic-solvent-based cleaners. Consequently, they have reduced their use of organic solvents by 30 percent. A management initiative, which rewards employees for ideas concerning waste reduction and cost savings projects, is also part of Hamilton Beach's program for pollution prevention.

ANNUAL COST SAVINGS: $20,260

PAYBACK PERIOD: 5 months

TIME TO IMPLEMENT: 3 months

TRANSFERABILITY: Using distillation to recover spent
                                solvent is a feasible alternative for
                                most degreasing solvents because of
                                their low boiling points. The quality
                                of the wastes generated may determine
                                whether recycling can be done in-house
                                or off-site. An off-site recovery
                                arrangement, like the one used by
                                Scovill, eliminates the capital
                                expenditures needed for distillation
                                equipment, and the maintenance and
                                oper- ating costs associated with such
                                equipment. The drawback to such an
                                arrangement is the liability the gen-
                                erator must assume, both for the mater-
                                ial during transport, and for the
                                burial of still bottom residues. Be-
                                cause off-site recycling can sometimes
                                result in contamination of recovered
                                solvent with a substance from another
                                user's process, it is to the genera-
                                tor's advantage to arrange to have
                                their solvent segregated during
                                distillation.

                                The substitution of other cleaning
                                agents for organic solvents is becom-
                                ing more commonplace, since the latter
                                are expensive, and present serious
                                health risks to workers.

                                Employee incentive programs are
                                also becoming popular, since many
                                waste reduction projects involve
                                changes in housekeeping practices, or
                                in equipment operation and maintenance;
                                opportunities for these modifications
                                are most readily noticed by production
                                personnel, rather than by management.

ENVIRONMENTAL AND
HEALTH BENEFITS: Reduced threat to air and groundwater
                                quality; conservation of resources;
                                improved health and safety conditions
                                for plant personnel.

Project Description

Scovill uses 1,1,1-trichloroethane solvent to degrease the metal stampings produced in their manufacturing process. Until implementation of their recycling program, the company had been accumulating waste solvent in 55-gallon drums. In order to avoid the responsibility and expense of landfilling the wastes, Scovill collaborated with the Fayetteville Office of the North Carolina Solid and Hazardous Waste Management Branch to develop a plan for recycling the waste solvent. Ashland Chemical Company was contracted to recycle 27 drums of already accumulated waste, as well as all future waste, which Scovill expects to generate at a rate of about 38,000 pounds per year.

A $3,000 storage building was constructed to house the wastes, and a central, recessed floor section was installed to minimize the pollution impact of an accidental spill. The structure can contain up to 220 gallons of spillage. Other miscellaneous equipment for the storage building includes sand, oil-dry, shovels, and fire extinguishers, at a total cost of about $250. Personnel and maintenance costs for the system run about $100 per year.

The company reports that substitution of the recycled solvent for the virgin product has reduced their overall raw material costs from $0.41 per pound to $0.27 per pound, yielding an annual savings of $5,320 per year. Scovill also eliminated all of their previous waste disposal costs, estimated to be about $0.08 per pound, or $3,040 per year. The result was an annual cost savings of $8,260, and a payback period of under 6 months. In addition to financial savings, the new storage building and the elimination of landfill requirements have reduced the potential for health and environmental risks. Although the firm is subject to RCRA cradle-to-grave accountability for their wastes, their reduction of waste volume has significantly reduced their liability risks.

Scovill also tested a water-soluble synthetic cleaner as a possible substitute for the 1,1,1-trichloroethane organic solvent degreaser. The cleaner is manufactured by Cincinnati Milacron Company of Cincinnati, Ohio. After some experimentation, they found the cleaner suitable for some of their applications, and have been able to reduce their 1,1,1-trichloroethane use by 30 percent. The water-soluble cleaner is not suitable for all applications because it can corrode galvanized parts. Scovill reports a $12,000 per year savings from this substitution.

Scovill has also instituted an employee incentive cost savings program which they hope will lead to other pollution prevention projects. Six-member teams are randomly selected, and the teams' membership is changed each year. The team responsible for the greatest annual cost savings to the company receives

151

bonus checks.   No waste management projects have resulted from this program yet, but Scovill is considering an in-house recycling program, which would be a project eligible for the bonus award.

PROCESS DIMENSIONS

===

WASTES GENERATED	Reduced by 38,000 pounds per year (less substitutions)
PRODUCTIVITY	No change
PRODUCT QUALITY	No change
DOWN TIME	No change
FACILITIES	$3,000 one-time cost
EQUIPMENT	$250 one-time cost
RAW MATERIALS	$17,320 saved per year
WATER	No change
ENERGY	No change
WASTE DISPOSAL	$3,040 saved per year
POLLUTION CONTROL	No change
PERSONNEL/MAINTENANCE	$100 cost per year
NET BENEFITS	Annual cost savings, resource conservation, and reduced risk of RCRA liability

===

Case Study 30

TYPE OF INDUSTRY: Manufacture of Electronic Telephone Switching
Equipment

NAME OF FIRM: GTE Sylvania
(Lancy International, consulting firm, William
McLay 412/452-9360)

LOCATION: Chicago, Illinois

SIC: 3661

```
***************************************************************
*                                                             *
*                      MODIFICATIONS                          *
*                                                             *
*  1) WASTEWATER TREATMENT SYSTEM CONVERTED TO CLOSED-LOOP    *
*                                                             *
*  2) INCORPORATION OF ELECTROLYTIC COPPER RECOVERY SYSTEM    *
*                                                             *
***************************************************************
```

Abstract

With stricter regulations concerning copper effluents soon to be enforced, GTE, aided by Lancy International, decided to modify the process water treatment of their copper-etching rinse stream by installing an electrolytic metal recovery cell. The electrolytic cell is part of a closed-loop system that follows the etching operations on the Electroless copper plating line. The cell was installed downstream from the rinse tank, where it recovers copper that would have ended up as a landfilled sludge. After the copper ions are removed, the rinse water is recirculated to the rinse tank. The copper is periodically removed from the cell's cathode and sold as No. 1 copper scrap. In its first year of operation this system saved $2,000, from the recycled copper scrap, and $4,000 from avoided waste management costs.

ANNUAL COST SAVINGS: $6,000

PAYBACK PERIOD: Information not provided.

TIME TO IMPLEMENT: Information not provided.

TRANSFERABILITY: Applicable to industries producing
                                    waste streams containing metal ions,
                                    especially copper.

ENVIRONMENTAL AND
HEALTH BENEFITS: Eliminated risks of surface- and
                                    groundwater contamination by hazardous
                                    wastes; conservation of water and
                                    metal resources; reduced chances of
                                    worker or non-worker contact with
                                    hazardous waste by reducing the
                                    formation of hazardous waste.

Project Description

In order to meet stricter copper-effluent regulations from
the Chicago Metropolitan Sanitary District, GTE investigated
opportunities to recover copper from their copper-etch
wastewater. They also looked for ways to reduce the plant's
water useage by recycling the rinse water from plating
operations. Through these modifications, GTE can meet the new
regulations, save on water costs, and recover valuable No. 1
copper scrap.

GTE recovers the copper through an electrolytic metal
recovery cell designed by Lancy Labs. The cell is part of a
closed-loop system that follows the etching operations on the
Electroless copper-plating line. The cell is installed downstream from the rinse tank, where it recovers copper that would
have ended up as a landfilled sludge. The now copperless rinse
water is recirculated to the rinse tank. The copper is
periodically removed from the cell's cathode and sold as No. 1
copper scrap. GTE saves $2,000 dollars through the sale of the
recovered copper, and saves $4,000 in avoided waste management
costs since the copper is now sold as a resource, instead of
being sent to a landfill as a waste.

The second of three waste streams, consisting of complexed
copper effluent from the Electroless plating operation, flows
into a tank, where sulfuric acid is added to destabilize the
copper complexes. The solution flows into the second tank, where
lime and sodium hydrosulfite are added to precipitate the copper.
The mixture is then filtered, and the pH adjusted, before final
release of the effluent to the sewer. The copper precipitate is
buried because, presently, the copper is not economically
recoverable.

The third waste stream, containing non-complexed copper-
bearing rinse waters and cleaners, flows into an equalization

tank to accomodate surges in the system, and then into a neutralization tank. In the neutralization tank, the copper is precipitated as copper hydroxide, which is passed into a third tank, where a polymer is added to accelerate the settling of the copper hydroxide. The pH is adjusted before the copper-free effluent is finally released into the sewer. The sludge is sent to a filter press for final dewatering.

Flow restrictors are one way of reducing the quantity of rinses; they have cut GTE's water use by 50 percent. The operating costs for the this system, including electricity, labor, and chemicals, are less than was originally anticipated.

PROCESS DIMENSIONS

==

WASTES GENERATED	Reduced CuOH sludge by 55 gallons per week
PRODUCTIVITY	No change
PRODUCT QUALITY	No change
DOWN TIME	No change
FACILITIES	None
EQUIPMENT	Investment, information not provided
RAW MATERIALS	75 pounds conserved per week
WATER	Reduced by 50 percent
ENERGY	Information not provided
WASTE DISPOSAL	Information not provided
POLLUTION CONTROL	Cost-effective
PERSONNEL/MAINTENANCE	One full-time operator
NET BENEFITS	Conservation and waste management efficiency maximized

==

For more information, see "Case History: Waste Water Treatment", in PC FAB, 1984.

Case Study 31

TYPE OF INDUSTRY: Manufacture of Printed Circuit Boards

NAME OF FIRM: Data General Corporation
                    (Stan Taylor 919/553-5076)
                    Clayton, North Carolina

SIC: 3573, 3679

MODIFICATIONS

1) IDENTIFICATION OF MARKETS UNTREATED WASTES

2) PROCESS CHANGES TO OBTAIN SLUDGES MARKETABLE FOR COPPER RECOVERY

3) LAND APPLICATION OF PRETREATED METAL-PLATING WASTES

Abstract

Data General Corporation generates several hazardous waste streams during the manufacture of printed circuit boards. Most of these wastes are characterized by high concentrations of heavy metals. In 1981, the company launched a vigorous program to reduce their reliance on hazardous waste landfills. A management initiative, to develop markets for their wastes, resulted in disposal cost savings of $9,000 per year, and in revenue from the sale of untreated wastes. Other wastes were treated with new process chemistry, and changes were made in equipment design or operation; these actions rendered additional material suitable for sale for metal recovery. A filtration system for process water yields an effluent clean enough for land application on company-owned forestland. Data General's program to produce saleable wastes, and to market those wastes, has reduced the hazardous material they landfill by over 400 tons per year, and has yielded a cost savings of $180,000 per year.

ANNUAL COST SAVINGS: $180,000

PAYBACK PERIOD: 1.5 months

TIME TO IMPLEMENT: 3 years

TRANSFERABILITY: The strategy of focusing on the sale of
                          wastes, at their current level of gener-
                          ation, rather than on reduction of waste
                          volumes, is feasible if the wastes con-
                          tain materials which are valuable, and
                          for which the recovery technology
                          exists. Use of a waste exchange might
                          help a generator to evaluate potential
                          markets, and to determine what
                          treatments might be necessary to make
                          wastes saleable. The use of a
                          consultant or assistance from a trade
                          association might also be helpful in
                          making these determinations. Some of
                          the treatment methods developed by Data
                          General resulted in significant
                          reduction of waste volumes, even though
                          this was not their primary objective.

                          Land application of treated plating
                          wastewater is not widely used; the
                          feasibility of such a system would
                          require significant engineering
                          evaluation, land availability, and will-
                          ingness to maintain a careful monitoring
                          program. Data General has shared data
                          from their operation with the North
                          Carolina Agricultural Extension Service,
                          which might also serve as a resource for
                          other firms considering this modi-
                          fication. The system's advantages are
                          that it reduces the pollution load on
                          municipal treatment facilities, and
                          allows the wastewater to serve as a
                          supplementary source of irrigation.

ENVIRONMENTAL AND
HEALTH BENEFITS: Reduced risk of soil and groundwater
                          pollution by heavy metals; conservation
                          of land and material resources. Because
                          of the potential for accumulation of
                          metals in the soil, frequent monitoring
                          of treated effluent, soil, plants, and
                          groundwater is required to ensure that a
                          land application system is functioning
                          properly.

159

Project Description

Data General Corporation, in Clayton, North Carolina, manufactures printed circuit boards for computers. The wastes from the fabrication, assembly, and testing of the circuit boards include spent etchant and plating baths; still bottom wastes from solvent cleaner reclamation; and process water contaminated with metals such as copper, nickel, and lead. Many of these wastes are hazardous and are generated in significant volume. In 1981, the firm set a goal -- landfill disposal should be used only as a method of last resort. They set out an ambitious plan of action, as follows:

1) All waste streams that have a monetary value, and that can be economically recovered, should be routed to some recovery process, on or off-site;

2) Remaining waste streams should be modified, by manufacturing or waste treatment process changes, so that they are marketable, or are no longer hazardous, by definition, and can be disposed of in the local landfill;

3) Wastes which cannot meet the above goals should be rendered suitable for disposal in a properly designed and licensed incinerator;

4) If all else fails, the remaining volume of waste should be reduced to a minimum and buried in an approved hazardous waste landfill.

This program required the evaluation of all phases of the Data General operation. The company points out that their strategy places primary emphasis on selling the spent chemicals and sludges, not on reducing or eliminating the production of these wastes. They believe that this strategy avoids the risk of sacrificing product quality or quantity for improved pollution control. They have modified certain manufacturing processes to serve the waste treatment program, but only with the full support of the process engineer. Data General reports that, over the last two years, approximately 15 percent of the total engineering effort in the printed circuit board fabricating process has been devoted to the waste management program. This effort has included marketing acitivities to find buyers for their wastes, operational revisions and development of new process chemistry, and several attempts to share their program with state personnel and university extension agents.

Sale of Untreated Wastes

Markets were identified for the following untreated waste streams:

1) Spent cupric chloride etchant

2) Spent plating baths
   a) acid copper bath
   b) palladium catalyst bath
   c) spent sulfuric acid from the etch-back process on the Electroless copper line

3) Copper sulfate crystals from regeneration of the sulfuric-peroxide bath

4) Spent ammonia etchant

5) Spent nickel-plating bath

6) Still bottoms from the 1-1-1 trichloroethane recovery process.

These wastes are shipped, either at no cost or for profit, to buyers, yielding the firm approximately $9,000 per year in disposal cost savings and revenue. The wastes total approximately 90,000 pounds per year.

Sale of Treated Wastes

Several wastes were treated to put them in a saleable form:

1) Spent Electroless copper bath is treated in a batch process for copper removal. By raising the pH of the bath and adding formaldehyde as a reducing agent, the copper in solution is destabilized. The bath is seeded with finely powdered copper sludge generated in a previous process, and the destabilized copper precipitates out as elemental copper. The resulting sludge is 90-percent metallic copper. This process will soon be revised to include the addition of copper-bearing spent acid baths, such as the sulfuric peroxide etchant, which will be added to the mixture above. The revised treatment operation will eliminate approximately 151,800 pounds of waste sludge per year at a savings of approximately $16,300 per year.

2) Process water, from printed circuit board fabrication, assembly, and testing, is treated using a cross-flow-filtration water purification system manufactured by Baker Brothers, Inc. The filters were physically modified so that the membrane filter bundles could be cleaned quickly and easily; the waste flow pattern and process chemistry were altered to eliminate the use of ferrous sulfate coagulant. The wastes are reduced with sodium

hydrosulfate and then precipitated out of solution. After the
ferrous sulfate is removed, the heavy-metal hydroxide sludge no
longer contains iron, and can be successfully processed for
copper recovery.

Waste streams, which are not treatable in the continous
water purification system (WPS), or which contain strong
chelating agents, or which are high in organic compounds that
interfere with the WPS, process are batch-treated using sodium
sulfide. The sulfide sludge is very fine, and careful control
of the process is necessary in order to produce a sludge which
can be removed in a filter press. Monitoring is also required to
prevent the formation of hydrogen sulfide and the resulting odor
problems. In order to maintain a high copper fraction in the
sludge, no coagulant, such as lime or ferrous sulfate, is used.
An organic polymer is used to help improve sludge
characteristics. The sulfide sludge is also often blended with
the metal hydroxide sludge from the WPS to improve filtration
characteristics during filter press dewatering. The batch
treatment process eliminates the shipment for landfill disposal
of approximately 18,000 gallons of sludge per year; this yields
about $16,500 per year in disposal cost savings and revenue.

Separation of waste streams and changes in process chemistry
decreased reagent costs by $74,000 per year, and cut sludge
volume in half. Transport and disposal costs, totalling $56,000
per year, were eliminated, and Data General receives revenue of
$7,200 per year from sale of the sludge. Equipment costs for the
modifications totalled $17,500.

Sludges, from the WPS and the sulfide batch precipitation,
are dewatered in a filter press. An air blow-down line was added
to the press at minimal cost; the solids concentration of the
filter cake has increased from 15 to 60 percent. Depth filters,
which were previously used to polish the press effluent, were
eliminated once the press operation was improved. The depth
filters had previously generated 300 pounds of spent filters per
month; these required landfill disposal.

On-Site Reuse of Treated Wastes

The copper-laden sludge from the process water filtration
system is treated and sold for copper recovery, as described
above. The treated process water is disposed of by application
to forestland owned by the company. This was the first site in
North Carolina for the land application of pretreated metal-
plating waste.

There are approximately 27 acres of land committed to the system. The company discharges about 200,000 gallons of treated water per day to the area; they are not permitted to deliver more than two inches of irrigation flow to the site per week. Copper is the limiting constituent of their wastewater, and their permit regulates the amount of copper they may apply.

Data General has a comprehensive monitoring program for the site. Surfacewaters above and below the system are sampled, and two monitoring wells were drilled for groundwater testing. The company has experienced no problems with the system, and anticipates that the site will remain suitable for cropland after the land application operation has ceased.

Incineration of Treated Wastes

Data General has begun to incinerate oil and flux waste from the wave folder operation in order to avoid land disposal of the hazardous lead-containing wastes.

In summary, Data General is aggressively pursuing its goal of eliminating the use of hazardous waste landfills. In the process, they have developed new methods: land application of pretreated metal-plating wastes; sulfide precipitation for strongly chelated wastes; and treatment of chelated process rinse waters without using conventional coagulants, which results in reduced sludge volumes. At the same time, they have reduced their disposal costs by over $80,000 per year. They estimate their total process cost savings, for the several modifications, to be more than $100,000 per year. Capital investment for the sludge recovery program was less than $50,000.

They are beginning pilot-testing of new equipment designed to electrolytically remove copper from solution. A woven fiber cathode will provide increased surface area for plating. The copper will be electrolytically stripped from the cathode onto a starter sheet. Because the system uses a captive electrolyte, no liquid waste solution is generated during the cathode stripping. The equipment is manufactured by Metal Removal Systems, a Division of PCK Technologies in New York. Data General anticipates that the process will be much more economical than the current system, and will produce a metal sheet which can be sold for scrap, rather than a copper sludge requiring further copper recovery operations.

PROCESS DIMENSIONS

===

WASTES GENERATED	Eliminated process water waste; reduced wastes requiring burial by 400 tons per year
PRODUCTIVITY	No change
PRODUCT QUALITY	No change
DOWN TIME	Decreased by 75 percent
FACILITIES	No change
EQUIPMENT	$50,000 one-time cost
RAW MATERIALS	Conserved, reduced costs
WATER	No change
ENERGY	No change
WASTE DISPOSAL	$80,000 saved per year
POLLUTION CONTROL	$100,000 saved per year
PERSONNEL/MAINTENANCE	No change
NET BENEFITS	Relief from hazardous waste management costs, annual cost savings

===

Case Study 32

TYPE OF INDUSTRY: Microelectronics

NAME   OF   FIRM: Minnesota Mining and Manufacturing (3M)
                    (Mike Koeningsberger 612/778-4523)

LOCATION: Columbia, Missouri

SIC: 367

```
*  ┌─────────────────────────────────────────────────────────┐  *
*  │                     MODIFICATION                          │  *
*  │                                                           │  *
*  │   REPLACEMENT OF A CHEMICAL PROCESS WITH A MECHANCIAL      │  *
*  │  PROCESS USING A ROTATING BRUSH FOR CLEANING METAL SHEETING│  *
*  │                                                           │  *
*  └─────────────────────────────────────────────────────────┘  *
```

Abstract

3M's microelectronics plant makes flexible electronic circuits from copper sheeting. Before use, the metal was cleaned by spraying it with ammonium persulfate, phosophoric acid, and sulfuric acid. This process resulted in the production of hazardous wastes that were expensive to manage properly. An alternative equipment design was introduced to replace the chemical cleaning process. The new process features a machine with rotating brushes that mechanically cleans the metal with pumice. The fine abrasive pumice leaves a sludge that is not hazardous and that can be placed in a conventional sanitary landfill. This new method reduced the plant's production of hazardous waste by 40,000 pounds a year. It also saved $15,000 in its first year, due to reduced raw material, disposal, and labor costs.

ANNUAL COST SAVINGS: Greater than $15,000

PAYBACK PERIOD: 3 years

TIME TO IMPLEMENT: Information not provided

TRANSFERABILITY: Applicable to industries manufacturing
                          printed circuit boards, and to those

165

                              with similarly high standards for
                              cleaning sheet metal.

ENVIRONMENTAL AND
HEALTH BENEFITS: Elimination of 40,000 pounds of
                              hazardous liquid waste per year, with
                              concurrent elimination of risks
                              associated with land disposal.

Project Description

     The copper sheet metal used by 3M in the manufacture of
flexible electronic circuits must be clean. In order to clean
the copper, it was sprayed with ammonium persulfate, phosphoric
acid; these chemicals clean the copper by dissolving residues or
contaminants. The solution which results from this cleaning
process must be managed as a hazardous waste.

     3M replaced the chemical treatment system with a specially
designed mechanical device that uses rotating brushes and an
abrasive pumice to clean the copper. The resultant pumice-metal
sludge is not hazardous and can be placed in a conventional
sanitary landfill. This process modification prevented the
production of over 40,000 pounds of hazardous liquid wastes per
year. It saved $15,000 in its first year, due to lower raw
material, labor, and disposal costs. Because of yearly
production increases, operational savings and prevented
pollution, volumes continue to rise.

PROCESS DIMENSIONS

WASTES GENERATED	eliminated the production of 40,000 pounds of hazardous waste per year
PRODUCTIVITY	No change
PRODUCT QUALITY	No change
DOWN TIME	No change
FACILITIES	No change
EQUIPMENT	$59,000 one-time cost
RAW MATERIALS	Cost reduced
WATER	No information provided
ENERGY	Need decreased
WASTE DISPOSAL	Hazardous waste management cost eliminated
POLLUTION CONTROL	Decreased cost
PERSONNEL/MAINTENANCE	Decreased cost
NET BENEFITS	Reduced production costs, improved resource conservation

Case Study 33

TYPE OF INDUSTRY: Printed Wiring Board Manufacture

NAME OF FIRM: Digital Equipment Corporation
                    (Lancy International, consulting firm, William
                    McLay 412/452-9360)

LOCATION: Tempe, Arizona

SIC: 3679

```
*~~~~~~~~~~~~~~~~~~~~~~~~~~~~~~~~~~~~~~~~~~~~~~~~~~~~~~~~~~~~~~~*
*                                                             *
*                       MODIFICATIONS                         *
*                                                             *
*   1) INSTALLATION OF PROCESS WATER TREATMENT SYSTEM WHICH    *
*      INCLUDES ELECTROLYTIC RECOVERY OF COPPER AND            *
*      ATMOSPHERIC EVAPORATOR RECOVERY OF CHROMIC ACID         *
*                                                             *
*   2) UTILIZATION OF OTHER INDUSTRIAL WASTES AS RAW MATERIALS *
*                                                             *
*~~~~~~~~~~~~~~~~~~~~~~~~~~~~~~~~~~~~~~~~~~~~~~~~~~~~~~~~~~~~~~~*
```

Abstract

Digital Equipment Corporation's new operation for manufacturing printed wiring board includes several resource conservation and recovery processes integrated within their process water treatment process. The system, designed by Lancy International, includes metal recovery by electrolytic cell and evaporation. The electrolytic reaction recovers 92 percent of the dissolved copper from dragout rinses, and the atmospheric evaporator recovers 95 percent of the chromic acid dragout. The chromic acid is recycled into the etch process line. The company also uses a lime slurry by-product, from a local acetylene gas manufacturer, as an acid neutralizer.

ANNUAL COST SAVINGS: Information not provided.

PAYBACK PERIOD: Information not provided.

TIME TO IMPLEMENT: Information not provided.

TRANSFERABILITY: Electrolytic recovery is applicable to
                        many metals, with varying success.

The composition of the wastewater
determines the effectiveness of the
technology. Evaporation is a simple
technology, easily applied given suitable conditions. In some areas,
atmospheric conditions are conducive;
another option is to recover waste
heat for evaporative energy.

ENVIRONMENTAL AND
HEALTH BENEFITS: Eliminated risks of surface- and
groundwater contamination by hazardous
wastes; conservation of water and
metal resources; reduced chances of
worker or non-worker contact with
hazardous waste.

Project Description

Digital Equipment Corporation's Tempe plant manufactures
printed wiring board. The management decided to optimize the
return on their effluent treatment investment by recovering the
copper that is usually lost in the effluent stream. This also
reduces chemical and sludge disposal costs.

Lancy International designed the centralized control system
which monitors the effluent streams from the plant's several
production processes. The system was installed 12 feet below the
production floor in order to utilize gravity flow for all
contaminated effluents; this eliminated the need for pumping the
wastewater to the treatment tanks. In order to save on chemical
costs, and to make use of a waste product from another industry,
acid wastes are neutralized by a lime slurry obtained from a
local acetylene gas manufacturer.

In order to make better use of Digital Equipment
Corporation's etching chemicals, Lancy International installed an
atmospheric evaporator. The evaporator recovers approximately 95
percent of the chromic acid dragout and recycles it into the
chromic acid etch line. The evaporator is also designed to
eliminate any atmospheric discharge.

PROCESS DIMENSIONS

===

WASTES GENERATED	Reduced
PRODUCTIVITY	No change
PRODUCT QUALITY	No change
DOWN TIME	No change
FACILITIES	Investment required, no details provided
EQUIPMENT	Investment required, no details provided
RAW MATERIALS	Copper and chromic acid conserved
WATER	Information not provided
ENERGY	Information not provided
WASTE DISPOSAL	Sludge disposal costs reduced
POLLUTION CONTROL	Greatly enhanced
PERSONNEL/MAINTENANCE	Information not provided
NET BENEFITS	Enhanced conservation profitablity

===

For more information, see Electronic Packaging and Production, May 1984, pp. 136-138.

Case Study 34

TYPE OF INDUSTRY: Metal Radiator Manufacture

NAME OF FIRM: Modine Manufacturing
                    (James Egide 414/636-1200)

LOCATION: Trenton, Missouri

SIC: 3714

```
*  ------------------------------------------------------------  *
*                                                                *
*                        MODIFICATION                            *
*                                                                *
*     INSTALLATION OF ION EXCHANGE AND ELECTROLYTIC              *
*            EQUIPMENT FOR COPPER RECOVERY                       *
*                                                                *
*  ------------------------------------------------------------  *
```

Abstract

Modine prepares brass parts for the manufacture of radiators by dipping them in a series of chemical solutions. The hydrogen peroxide/sulfuric acid bright dip was causing them several problems; these included the buildup of copper sulfate pentahydrate crystals that were being deposited in a hazardous waste landfill. There was, in addition, a significant dragout of copper ions into the subsequent rinse tanks. Rather than incur the initial and long-term landfilling costs for the hazardous wastes, Modine elected to eliminate the formation of the crystals by continuously recirculating the hydrogen peroxide/sulfuric acid bright dip solution through an ion exchange column. The column selectively removes the metal. The purified dip solution is recycled to the bath. Once the ion exchange resin becomes saturated, the copper ions are exchanged again in the regeneration process, and are subjected to an electrolytic reaction to remove them from the solution. The plated copper is sold as No. 1 copper scrap.

ANNUAL COST SAVINGS: $22,000

PAYBACK PERIOD: 14 months

TIME TO IMPLEMENT: 8 months

TRANSFERABILITY:	Applicable to industries producing waste streams containing elements of sufficient value and in sufficient quantity to merit their recovery. Ion exchange and electroplating efficiencies will vary with different elements. Economics indicate that valuable metals provide the best returns, especially when their recovery will reduce waste management costs. Implementing this process during the design phase of a system yields the best economic efficiency, but it can also be included in the modification of an existing facility.
ENVIRONMENTAL AND HEALTH BENEFIT:	Eliminated risk of surface- and groundwater contamination by hazardous wastes; conservation of metal resources; eliminated risks of worker or nonworker contact with hazardous waste by preventing formation of the waste.

Project Description

At Modine's Trenton, Missouri plant, brass and copper sheet metal radiators are built. In the radiator manufacturing process, sheet brass is first formed into the required variety of parts. The parts are cleaned with an aqueous alkali degreaser (for removal of drawing lubricants), followed by a pair of counterflowing rinses and a hydrogen peroxide/sulfuric acid bright dip. The bright dip removes surface oxides from the brass parts. The parts are then dipped again into two counterflow rinses, followed by the application of a low-molecular-weight anti-oxidant coating. These steps are required in order to ensure a clean surface for subsequent soldering and brazing operations.

In the bright dipping process, considerable amounts of copper and zinc metal are etched from the surface of the parts being processed. As a result, copper sulfate crystals accumulate in the bottom of the tank and require periodic removal and disposal. A significant amount of these divalent cations is dragged into the subsequent rinse tank. In order to avoid incurring the high costs of disposing of these wastes, and to avoid contaminating rinse waters, Modine elected to eliminate the formation of the crystals.

172

An ion exchange column and a plating cell were added to the system. The hydrogen peroxide/sulfuric acid bright dip solution is continuously recirculated through the ion exchange column. The column selectively removes the copper and returns the purified dip solution to the bath. Once the ion exchange column becomes saturated with copper, it is regenerated; the copper ions are fed into an electroplating cell, where the copper is plated out. The copper is then returned to Modine's suppliers for credit as No. 1 scrap.

PROCESS DIMENSIONS

WASTES GENERATED	Eliminated hazardous wastes
PRODUCTIVITY	Moderate increase
PRODUCT QUALITY	No change
DOWN TIME	Moderate reduction
FACILITIES	No change
EQUIPMENT	$27,000 one-time cost
RAW MATERIALS	$22,000 per year saved
WATER	Information not provided
ENERGY	Moderate increase
WASTE DISPOSAL	Eliminated disposal costs
POLLUTION CONTROL	Reduced cost
PERSONNEL/MAINTENANCE	Moderate increase
NET BENEFITS	Savings through resource conservation and waste reduction, reduced opportunity for environmental insult

Case Study 35

TYPE OF INDUSTRY: Lamination and Coating of Specialized Paper, Film, and Foil Products

NAME OF FIRM: Rexham Corporation
(Gordon Miller 704/847-9171)

LOCATION: Matthews, North Carolina

SIC: 3861

MODIFICATIONS

1) DEVELOPMENT OF A COMPREHENSIVE MANAGEMENT PROGRAM TO REDUCE WASTES REQUIRING LANDFILL DISPOSAL

2) SEGREGATION OF SPENT CLEANING SOLVENTS FOR REUSE

3) RECOVERY AND SALE OF SOLVENT VAPORS COLLECTED BY CARBON ADSORPTION

4) OFF-SITE DISTILLATION OF WASTE SOLVENT FOR REUSE

5) SOLVENT INCINERATION FOR HEAT RECOVERY

Abstract

The Rexham Corporation facility in Matthews, North Carolina, laminates and coats specialized paper, film, and foil substrates for the photography industry. A management initiative, to develop a comprehensive waste management plan for the firm, has resulted in several pollution prevention modifications. The initiation of new housekeeping practices, which segregate spent toluene solvent by ink color and type, has resulted in almost 100-percent reuse of liquid toluene. Toluene-laden vapors are collected in a carbon-adsorption-based air pollution control system, and are sold for use in the coatings industry. Off-site distillation of spent solvent mixtures has allowed the reuse of methyl-ethyl-ketone solvent for washup operations. In addition, Methyl-ethyl-ketone vapors are removed from the air and incinerated for heat recovery.

175

ANNUAL COST SAVINGS:	Information not provided.
PAYBACK PERIOD:	Information not provided.
TIME TO IMPLEMENT:	Installed during facility construction.
TRANSFERABILITY:	Organic vapor recovery is an expensive modification; payback may be realized in terms of improved air quality and personnel safety, rather than annual cost savings. New systems are now available that do not require the personnel commitments that the Rexham unit did. Use of a carbon adsorption system would depend on the retentivity of the solvent vapors on the carbon particles, the amount of vapor generated, and the capacities of the systems available.

Segregating spent cleaning solvents is particularly applicable when it allows the reuse of the reclaimed material in-house, as it has at the Rexham plant. Even if there is no other in-house use for such materials, segregation of waste streams may result in material suitable for use in other firms. Participation in a waste exchange might help identify other users and reduce disposal costs. Off-site solvent recovery by distillation can significantly reduce raw material costs, as well as waste disposal costs and risks. Several vendors in North Carolina provide such services.

Removing solvent vapors from air and incinerating them is an excellent way to reduce health risks to workers and reduce operating costs. The requirements for such a system are a sufficient concentration of solvent vapors and a use for the heat generated. |
| ENVIRONMENTAL AND HEALTH BENEFITS: | Improved air quality, reduced reliance on landfill disposal, conservation of |

energy and material resources, improved
health and safety conditions for plant
personnel.

Project Description

The Rexham Corporation facility in Matthews, North Carolina
is part of the company's Industrial Division. This facility
laminates and coats paper, film, and foil products, and prints
film substrates for the photography industry. Waste streams are
generated from several routine operations, as well as from
intermittent or one-time-only services for customers.

Rexham has made a firm commitment to waste reduction
activities. They view good hazardous waste management practices
as a way to maintain environmental integrity, to ensure worker
safety, and to increase job opportunities and profitability. The
Industrial Division has developed an on-going comprehensive
program to review and improve safety and waste management
practices throughout their operations. They hold an annual
meeting for representatives from the several division facilities.
These meetings provide an opportunity to review safety
contingency plans and hazardous waste handling practices, and to
discuss waste management problems of mutual concern.

At the Matthews plant, a Safety and Environmental Affairs
Manager oversees a number of activities to monitor and reduce
waste generation. A hazardous-waste manual has been developed
for plant personnel. The manual sets forth Rexham's contingency
plan for accidents and spills, and identifies the response teams'
activities. The plant's major waste streams are also reviewed in
the manual, including a description of what constituents are
present in each stream, what constituents are not permitted in a
given stream, and how each stream is to be handled. All waste
waste materials are tracked through the system to ensure proper
handling, and to identify possible problems.

Because much of their work involves on-going, product
development research, the variety of chemicals used at the
Matthews plant is very large -- in excess of 1,300 unique
chemicals. Management procedures call for the Safety and
Environmental Affairs Managers' approval on all incoming
materials. This practice permits a review of all new chemicals,
and allows special waste handling arrangements to be made, if
necessary. Periodic internal exchanges of surplus chemicals
between product development groups have also been organized; in
this way, excess chemicals are not wasted or allowed to become
outdated.

Rexham reports four major pollution prevention projects that have contributed to the success of their waste management program:

- o segregation of spent toluene solvent to allow reuse
- o recovery and sale of toluene vapors
- o off-site recovery of methyl-ethyl-ketone from spent solvent mixtures
- o incineration of waste solvent vapors for heat recovery

The capital improvements associated with these projects were made during construction of an adjacent manufacturing facility.

Liquid toluene is used for print and roller cleanup on various printing lines. By segregating the used toluene according to the color and type of the ink contaminant, each batch can be reused for thinning the same ink type and color. This system has allowed almost 100-percent reuse of liquid toluene.

Use of toluene solvent in another operation generates large quantities of toluene solvent vapors. In 1979, Rexham installed a carbon-adsorption recovery unit to capture these vapors. The carbon beds are regenerated in-house to remove the trapped toluene. The toluene is then sold to, the coatings industry for use as a thinner. The solvent recovery units achieve about 90-percent recovery; Rexham sells the reclaimed solvent for about half the price of virgin solvent.

The solvent recovery unit was a major capital investment and requires a full-time operator. A supervisor and three staff personnel are employed to oversee the operation. The carbon bed will eventually require complete replacement, but this will be an infrequent expense. Rexham reports a small financial return of 1 to 1.5 percent per year from the system, but the major benefits are improved air quality and increased opportunity for Rexham to expand its operations. Before the recovery system went on-line, the company released up to 5,000 tons of hydrocarbons into the atmosphere each year. The carbon adsorption system has reduced these emissions tenfold, to only 485 tons per year.

Another major waste stream at Rexham's Matthews facility is a solvent mix in which methyl-ethyl-ketone usually predominates. About 700 drums of such wastes are produced each year; they are sent off-site for distillation recovery of methyl-ethyl-ketone. Rexham currently contracts with Seaboard Chemical Company in Jamestown, North Carolina, for this processing; approximately 60- to 65-percent recovery is achieved. Rexham uses the reclaimed solvent for washup operations and Seaboard incinerates the still bottoms. Rexham is currently looking at in-house distillation as a way to reduce their waste management costs, and to perhaps achieve improved recovery efficiency.

Air, containing methyl-ethyl-ketone vapors, is vented through a continuous-flow, REECO retherm unit for heat recovery. The air/vapor mixture is passed through columns filled with a refractory material; here, the solvent is preheated and then incinerated. The cleaned air is pulled through the system and discharged, through a stack, to the ambient air. The heat produced is used in one of Rexham's drying operations.

Rexham recently established weekly, one-hour departmental meetings; thus providing a new opportunity for identifying and implementing improved waste management practices. The group meetings were instituted so that employees could participate in solving some of the work-related problems that affect them. The groups address a variety of topics, including ideas for improving safety practices, product quality, waste management practices, and office procedures.

PROCESS DIMENSIONS

===

WASTES GENERATED	Reduced air emissions (4,515 tons per year collected in carbon adsorption system)
PRODUCTIVITY	Increased plant capacity was possible with new air pollution controls
PRODUCT QUALITY	No change
DOWN TIME	No change
FACILITIES	Information not provided
EQUIPMENT	Information not provided
RAW MATERIALS	Reduced due to recycling and reuse
WATER	Information not provided
ENERGY	Reduced due to heat recovery
WASTE DISPOSAL	Reduced due to recycling and reuse
POLLUTION CONTROL	Increased cost (solvent recovery unit)
PERSONNEL/MAINTENANCE	3 employees for solvent recovery unit

```
-----------------------------------------------------------------
NET BENEFITS                        Annual cost savings,
                                    opportunity for plant
                                    expansion, decrease in worker
                                    exposure to solvents,
                                    significant improvement in air
                                    quality
=================================================================
```

Case Study 36

TYPE OF INDUSTRY: Electric Utility
NAME OF FIRM: Carolina Power and Light Company
(Grover Dobbins 919/836-6083)
LOCATION: Shearon Harris Energy & Environmental Center
Route 1, Box 327
New Hill, North Carolina 27562
SIC: 4911

```
*  --------------------------------------------------------  *
*                                                            *
*                      MODIFICATION                          *
*                                                            *
*          SALE OF FLY ASH AND BOTTOM ASH FOR REUSE          *
*                                                            *
*  --------------------------------------------------------  *
```

Abstract

Carolina Power and Light Company (CP&L) has been selling waste fly ash and bottom ash for reuse since 1973. The Company does not sell ash directly to users, but has contracted with another firm, Monier Resources, Inc., to oversee the marketing of the ash. Prior to this arrangement, all of the ash produced was sluiced to on-site ash ponds. The Company has now reduced its disposal costs and prevented the burial of useable material.

ANNUAL COST SAVINGS: Variable

PAYBACK PERIOD: Unknown

TIME TO IMPLEMENT: Information not provided.

TRANSFERABILITY: The success of fly ash and bottom ash sale depends on several factors, including the composition of the ash, the local market options, and the current costs of ash disposal. Ash composition can be affected by the type of coal burned, boiler design, operating conditions, and the methods used to handle and collect the wastes. Utility companies have a large resource base from which to evaluate and develop marketing programs. Any firm producing significant quantities of waste ash could pursue opportunities for selling or donating the ash in order to reduce the costs and risks of storage and disposal. The Electric Power Research Institute has prepared a by-product utilization manual which contains a step-by-step methodology for power companies to estimate the feasibility of marketing ash. Other industry groups may also be able to use this information; or they may need to investigate material composition, alternate uses, and available markets on their own, or with the assistance of their respective trade associations.

ENVIRONMENTAL AND
HEALTH BENEFITS: Conservation of land and material resources, and of energy used to produce materials for which fly ash can be substituted; reduced opportunity for accidents related to fly ash ponds; reduced threat of surface- and groundwater contamination from ash pond effluent.

Project Description

The Carolina Power and Light Company's coal-fired power plants currently produce 800,000 to 900,000 tons of fly ash and 300,000 to 350,000 cubic yards of bottom ash annually. This ash is typically sluiced to on-site ash ponds. Since 1972, CP&L has been selling limited quantities of ash. They are currently working with Monier Resources, Inc. (MRI) to expand their ash utilization activities. CP&L does not sell ash directly to users, but works through MRI, which purchases the ash from CP&L and manages its resale to users.

Initially, the fly ash and bottom ash sold were from the Roxboro Steam Electric Plant. Subsequently, bottom ash has been sold from the Sutton Steam Electric Plant (from 1979 to 1982); fly ash sales from the Mayo Electric Generating Plant began when Unit No. 1 started commercial operation in 1983. Bottom ash sales from the Mayo plant began in 1984.

Fly ash must be dry-handled in order to be saleable. Fly ash, collected by electrostatic precipitators, is transferred to a storage silo, from which it is loaded into haul vehicles (trucks or rail cars) for delivery to users. Producing saleable bottom ash typically involves wet-transfer of the ash to a reclamation area, to allow drainage, before it is loaded for transport.

The amount of ash sold is variable from year to year, depending on market conditions. Whatever the volume of ash sold, it reduces the land required for disposal, conserves the raw materials and energy required to produce it, and reduces the potential for pollution from ash ponds. The following is a summary of recent ash production and sales volumes:

Plant	Type of Ash	Production (units in thousands)	Sale (units in thousands)	
Roxboro	fly	400-500 tons	30-40 tons	(5-10%)
	bottom	150-200 cu. yds.	5-45 cu. yds.	(20-30%)
Sutton	bottom	30-35 cu. yds.	4 cu. yds.	(10-15%)
Mayo	fly	150-175 tons	50-60 tons	(30-35%)
	bottom	55-65 cu. yds.	no data available	

The primary equipment investment is reported to be highly site-specific. The investment is more quickly recouped when the system is installed as the sole ash-handling process, rather than when it duplicates the dry process alongside the wet process. A rough estimate of construction expenses, offered by the CP&L representative, is $1 to $2 per ton of ash, based on designed load; equipment operation and maintenance expenses are estimated to be $3 to $4 for each ton of ash generated per year.

PROCESS DIMENSIONS

==

WASTES GENERATED	Reduced by up to 30 percent
PRODUCTIVITY	No change
PRODUCT QUALITY	No change
DOWN TIME	No change
FACILITIES	Information not provided
EQUIPMENT	Information not provided
RAW MATERIALS	No change
WATER	Information not provided
ENERGY	Information not provided
WASTE DISPOSAL	Information not provided
POLLUTION CONTROL	Information not provided
PERSONNEL/MAINTENANCE	Information not provided
NET BENEFITS	Reduced waste management costs and risks

==

Case Study 37

TYPE OF INDUSTRY: Electric Utility

NAME OF FIRM: Duke Power Company
                   (David Roche 704/373-8775 and
                   Russell Propst 704/373-2377)

LOCATION: Charlotte, North Carolina

SIC: 4911

```
**********************************************************
*                                                        *
*                     MODIFICATIONS                      *
*                                                        *
*    1) SALE OF FLY AND BOTTOM ASH                       *
*                                                        *
*    2) INCINERATION OF POLYCHLORINATED BIPHENYLS (PCBs) *
*                                                        *
*    3) SEGREGATION OF WASTE STREAMS TO REDUCE           *
*       HAZARDOUS WASTE VOLUME                           *
*                                                        *
*    4) DECONTAMINATION AND REUSE OF RADIOACTIVE TOOLS   *
*       AND EQUIPMENT                                    *
*                                                        *
*    5) REVISED EQUIPMENT OPERATION TO IMPROVE WASTE     *
*       TREATMENT EFFICIENCY                             *
*                                                        *
*    6) A MANAGEMENT INITIATIVE TO ESTABLISH AN          *
*       ON-GOING PROGRAM FOR WASTE MINIMIZATION          *
*                                                        *
**********************************************************
```

Abstract

Duke Power Company has reduced waste production in three major areas of operation. Fly ash and bottom ash wastes from coal-fired plants are now offered for reuse as a concrete constituent, plastics component, or roadway construction material. Mineral oil insulating fluid contaminated with the organic chemical, polychlorinated biphenyl (PCB), is burned, as a supplemental fuel for at a steam electric generating station, for heat recovery. The Company has also instituted a comprehensive program for reducing low-level radioactive wastes (LLRW). The program includes: housekeeping improvements; the purchase of equipment to allow cleaning rather than disposal of contaminated tools and equipment; conversion from an evaporator-based processing system for non-recyclable waste to one based on ion exchange equipment; redesign of existing evaporator equipment so that it can be used to process recyclable waste; and, review and revision of filtration equipment operation in order to improve its performance, and to minimize replacement and disposal costs. The Company attributes the success of these modifications to the team approach encouraged by their management initiative.

ANNUAL COST SAVINGS: $6,184,000

PAYBACK PERIOD: Less than 1 year

TIME TO IMPLEMENT: 3 years (radwaste program). Complete information on the implementation times for other modifications was not provided.

TRANSFERABILITY: All of the programs and modifications described here would obviously be applicable to other companies engaged in coal- and nuclear-based energy production. However, any firm producing significant quantities of waste ash could pursue opportunities for selling or donating the ash in order to reduce their storage and disposal costs and risks. PCB destruction by incineration is an excellent method of disposal, but it may not be cost-effective for firms producing only small quantities of PCB-contaminated material. An out-of-house or cooperative arrangement might be feasible for these small firms.

Duke Power's management policies for handling low-level radioactive wastes can be adopted by any firm wishing to institutionalize the pollution prevention philosophy. Regardless of what waste products are generated, giving employees time and encouragement to review existing processes, and to propose improvements will reveal opportunities for waste reduction. Housekeeping improvements involving

the separation of hazardous and
non-hazardous waste streams,
reviewing cleaning procedures, and
adopting a policy of vigilance and
cooperation are all methods that can
certainly be adapted to any
industrial workplace.

ENVIRONMENTAL AND
HEALTH BENEFITS:
Reduced risk of PCB and low-level
radioactive waste contamination of
soil, surface- and groundwater; reduced risk of PCB bioaccumulation;
reduced risk of hazardous waste
transportation accidents; conservation of land and material
resources.

Project Description

Duke Power Company is one of the major electric power
generating firms in North Carolina. They have implemented
several important pollution prevention activities; these have
brought them annual cost savings, and have served as an example
to other industries in the state. Duke Power has reduced waste
production in three major areas of their operation: fly ash and
bottom ash wastes from their coal-fired power plants, low-level
radioactive wastes from their nuclear power plants, and PCB
wastes from their electrical distribution equipment.

Fly Ash and Bottom Ash Wastes

Since the late 1960s, Duke Power has sold some of the fly
ash from one or more of their coal-fired power plants; the ash is
used as a component in portland cement for producing concrete or
as a filler in asphalt. Electrostatic precipitators remove the
ash from stack gases. The ash is then conveyed to a storage silo
where it can be loaded onto trucks. The amounts of fly ash sold
for concrete products from Duke's several facilities were:

o Marshall Steam Station, 1979 -- 116,700 tons
o Cliffside Steam Station, 1983 -- 79,400 tons
o Belews Creek Steam Station, 1983 -- 42,700 tons

Selling fly ash generates revenue and avoids the costs of
sluicing and ponding the fly ash for disposal. In some
instances, ponded fly ash (cenospheres) has been removed from the

ash ponds and sold, in small quantities, to a plastic manufacturer for use as a filler.

Because fly ash constitutes at least 80 percent of the ash produced by burning coal, Duke Power is seeking to expand fly ash reuse to markets that do not have the quality control restrictions of concrete production. The pozzolanic property, light unit-weight, and compressive strength of fly ash make it desirable as a structural backfill, and as a roadbase stabilizer. This year, using 1,200 tons of fly ash, Duke Power contractors have constructed a one-mile road at the Marshall Steam Station; the road demonstrates stabilization techniques that the North Carolina Department of Transportation may be able to adopt in their roadbuilding and repair projects. If large volumes of fly ash could be used in road construction, disposal costs for the ponding or landfilling of this material would be greatly reduced or totally eliminated.

Since 1979, bottom ash from the Belews Creek Station has been sold for use as lightweight aggregate in concrete production. In 1981, 65,400 tons were sold. Since 1983, bottom ash from the Marshall Steam Station has been provided, without cost, to the North Carolina Department of Transportation; it is used, in place of crushed stone, as an aggregate base-course applied to unpaved secondary roads. The state has used approximately 30,000 tons of bottom ash in this way, as of October 1984. Bottom ash is sluiced from the bottom of the steam boiler to holding ponds, where it is dewatered and then reused. The periodic excavation of ash from the disposal ponds saves storage space, and minimizes the need for making other disposal arrangements.

PCB Disposal

Duke Power's efforts to reduce the amount of PCB wastes requiring hazardous waste disposal are part of a comprehensive program, across its service area to phase out PCB contamination from their electrical distribution system.

When electrical equipment in Duke's distribution system needs repair the electrical unit's fluid is tested for PCB concentration. If the fluid contains no PCB, it is reprocessed at a company facility, and used again; this saves $400,000 annually, compared to the cost of new mineral oil. If the fluid is contaminated with the regulated levels of PCB, it is burned as a supplemental fuel (the heat value is identical to that of fuel oil) in a high-efficiency boiler (Riverbend Steam Station). This incineration is approved by the Environmental Protection Agency, and has been proven to totally destroy PCB, as is shown by Duke's own stack emission tests. The contaminated oil, by being used for heat recovery at a steam electric generating station, does not place additional demand on the limited capacity of commercial hazardous waste disposal facilities. The mineral oil storage and feed system at Riverbend required approximately $75,000 in capital investment, but the system has saved Duke Power $168,000 in disposal costs since July 1981. The system, and PCB management program as a whole, are expected to continue realizing substantial economic benefits, while also removing PCB from the electrical distribution network.

Radioactive Waste Reduction

Because all processed low-level radioactive wastes must be packaged, shipped, and disposed of on a restricted-volume-basis in a hazardous waste facility, or must be released to the environment in controlled, dilute quantities, it was imperative that Duke Power minimize the production of these wastes at their new McGuire Nuclear Station. A team of staff members, from the Company's Design Engineering and Nuclear Production Departments, was assembled to aggressively pursue waste reduction strategies. The low-level radioactive solid wastes studied included:

a) Contaminated compactible materials: paper, plastic, and other materials, for both office work and radiation protection, used in the traditionally viewed "radiation area" of the plant. The projected volume was 30,000 cubic feet per year.

b) Contaminated tools and equipment: maintenance tools used on radioactively contaminated components, and components removed for replacement. The projected volume was 19,000 cubic feet per year.

c) Evaporator concentrate: the concentrated by-product of low-level radioactive liquid waste purification. The projected volume was 11,000 cubic feet per year.

Contaminated compactible materials were segregated by area within the plant into "potentially contaminated" and "not contaminated". A designated area and equipment were set up to survey all materials from non-contaminated collection points in order to give final assurance of radioactive cleanliness. In this way the quantities of material were substantially less than the volumes normally generated in similar nuclear plants. They were also in compliance with regulations and industry "good practices" standards. For the contaminated wastes, improved compactor equipment was purchased, installed, and placed in operation. This equipment further reduced the volume of wastes requiring disposal.

190

It was determined that contaminated tool and equipment volumes were higher than those that could be achieved by using the most effective technology for decontamination (removal of activity from the surfaces such that the component is no longer a source of radioactivity). The station issued contracts to a variety of service companies to process these components using a variety of decontamination techniques. The program allowed the clean-up of components such that disposal was no longer necessary, and allowed the collection of data on the effectiveness of each technique. The data were used to choose the most efficient equipment for purchase and permanent installation in 1984. Capital costs totalled $200,000 and the modifications have saved Duke Power approximately $1 millon per year.

The station reviewed the production of evaporator concentrates and questioned its necessity in relation to the newer approach of processing non-recyclable liquid wastes using ion exchange technology. Laboratory testing and review of early nuclear industry data were conducted. Existing systems for processing non-recyclable wastes were converted from evaporator-based to ion-exchange-based. The objective was to eliminate production of evaporator concentrate while also generating less ion-exchange resin by-product. The ion exchange modifications included the capability to supplement permanent equipment with portable equipment. Processing equipment can now be adapted to changing waste properties so that effluents released to the environment remain consistently low.

Ion exchange processing of non-recyclable liquid waste processing represents the application of a technology used only infrequently until recently. The success of this technology at McGuire placed the station and Duke Power Company among the early leaders in a trend toward the more widespread use of this process in the nuclear industry.

Low-level radioactive liquid wastes were handled by:

a) Evaporative processing. Using evaporators to treat non-recyclable wastes from equipment drainage and cleaning was difficult given the type of evaporator provided by the supplier of plant equipment. Evaporator operation demonstrated lower than designed process capacity, and produced large volumes (11,000 to 20,000 cubic feet per year) of concentrate as a by-product. Reduced process capacity forced the processing of recyclable liquids by other systems, resulting in increased release of dilute quantities of boron and tritium to the environment. The systems, as originally installed, lacked the capability to adapt to changes in waste stream chemistry and radionuclide concentrations.

b) Filtration equipment. During preoperational tests, the size and capacity of the filtration equipment were discovered to be inadequate for routine processing. A throughput rate was less than 4,000 gallons per cartridge changeout. The cost for replacing the spent cartridges was greater than $200,000 per year.

It was determined that the existing evaporator equipment was better suited for processing recyclable liquid wastes. Evaporator performance problems at other plants indicated that, unless the evaporators were improved or replaced, recyclable waste backlogs would require drainage of this waste to the non-recyclable waste process streams. A test program was conducted, and the evaporators extensively modified. The evaporator originally designed for non-recyclable waste processing was converted to process recycleable wastes so as to minimize boron and tritium release to the environment, and to minimize process costs. (This conversion was allowable due to the ion exchange modification of the non-recyclable wastes process.) Evaporator performance increased from 60 percent to 110 percent of the designed process rate. The combined, ion exchange and evaporator modifications cost $1,030,000 and have yielded an annual payback of $2 million. The evaporators were converted to recyclable waste processing at a time when many utilities were abandoning their use. This conversion placed Duke Power among industry's leaders in evaluating and utilizing unit processes based on waste stream and equipment compatibility.

Filtration equipment was reviewed, tested, and modified to improve filter performance and to minimize replacement and disposal costs. The modifications required a $60,000 investment, but they have saved $2 million per year. Operating data show that the rate of radioactive cobalt removal doubled, and that throughput to exhaustion improved by a factor of 25.

These programs and modifications were in place by the end of 1982. During 1983, both McGuire reactor units were either operated or tested under conditions which produced radioactive by-products. Program results through September 30, 1983, showed low-level radioactive waste volumes from McGuire to be less than 15 percent of the industry average, despite outages for major component modifications (high-volume waste generation) during this period. No evaporator concentrates were generated as a waste by-product. Resin from ion-exchange processing produced less than 5 percent of the projected 11,000 cubic feet of concentrates. Boron releases were reduced by 54 percent and tritium releases were reduced by 60 percent due to recycleable waste processing. Filtration cost was reduced by a factor of 24. Radioactive concentrations in released liquids were below the

192

levels permitted under the station license and federal regulations by a factor of greater than 100.

Duke Power Company has encouraged improved waste management by creating or expanding waste management groups at each of its stations. At McGuire Nuclear Station (MNS), a special Radioactive Waste Processing Introduction in the McGuire Chemical Manual stresses the importance of:

a) "participation of all plant personnel in notifying the radwaste organization of unusual occurrences such as leakage of coolant or oil; segregating waste streams where feasible; and minimizing extraneous waste such as cleaning chemicals and trash from sumps and floor drains."

b) "coordination and cooperation in reduction of waste production; efficient scheduling and performance of system maintenance; timely waste sampling, analysis, and receipt of results and recommendations; and system operation to the highest professional standards." The policy statement concludes, "The most fundamental concept of MNS radwaste philosophy is waste release minimization through operational optimization. This concept should be distinctly segregated from the notion of releasing radioactive waste at a rate which merely satisfies a pre-determined legal limit."

A cost-benefit analysis of the improvements made in low-level radioactive waste programs and equipment showed that net savings were realized within one year of their implementation.

Duke has also identified several other benefits of the program:

1) Use of the waste disposal site has been minimized through more efficient, effective control and processing of solid and liquid wastes;

2) The number of low-level radioactive waste shipments has been reduced;

3) Reclamation and reuse of recyclable liquid waste reduced release of boron and tritium to the environment;

4) Liquid releases from the station are maintained at a small fraction of permissible levels.

193

PROCESS DIMENSIONS

==

WASTES GENERATED	Waste fly ash reduced by approximately 250,000 tons per year
	Waste bottom ash reduced by approximately 95,000 tons per year
	Volume of PCB waste reduced
	Volume of Solid LLRW reduced
	Projected volume of LLRW evaporator concentrates reduced by 95 percent
	Tritium and boron release reduced by 60 percent
PRODUCTIVITY	No change
PRODUCT QUALITY	No change
DOWN TIME	No change
FACILITIES	No change
EQUIPMENT	$1,290,000 one-time cost (LLRW)
	$75,000 one-time cost (PCB incineration)
RAW MATERIALS	$400,000 saved per year (insulating fluid recovery)
WATER	Information not provided
ENERGY	Information not provided

WASTE DISPOSAL	Approximately $2,500,000 saved per year (fly ash)
	$84,000 saved per year (PCB)
	$3,200,000 saved per year (solid LLRW)
POLLUTION CONTROL	Information not provided
PERSONNEL/MAINTENANCE	Information not provided
NET BENEFITS	Annual cost savings, extended landfill life, reduced environmental and health risks

Case Study 38

TYPE OF INDUSTRY: Solid Waste Landfill

NAME OF FIRM: Natural Power (Bill Roland 919/876-6722)

LOCATION: Raleigh, North Carolina

SIC: 4953

```
***************************************************************
*                                                             *
*                      MODIFICATION                           *
*                                                             *
*         INSTALLATION OF A BIOGAS RECOVERY                   *
*                SYSTEM AT A LANDFILL                         *
*                                                             *
***************************************************************
```

Abstract

Natural Power, operating from the Gresham Lake landfill in Raleigh, North Carolina, has implemented a methane recovery system for converting "biogas" to electric power, which is then sold to the local utility. Natural Power recovers combustible gas generated from the anaerobic decomposition of municipal wastes buried at the landfill. The gas is collected, under vacuum, from a number of wells sunk into the landfill. After filtration to remove particulates, tha gas is fed directly into two diesel generators modified to use a methane fuel source. The electricicty generated at the landfill is metered and fed into the local electric utility's transmission lines.

ANNUAL COST SAVINGS: $25,700

PAYBACK PERIOD: Variable, dependent on the rate of methane recovery and the price per kilowatt-hour (kWh) paid by utility.

TIME TO IMPLEMENT: Estimated 6 months to 3 years.

TRANSFERABILITY: Applicable at any landfill which has received organic solid waste and is generating gaseous by-products of anaerobic decomposition. Variables affecting the economics of such a facility include cost to produce and price received per kWh from the local utility, time which has elapsed since landfill was closed, size of landfill, production rate of gas, and percent methane composition of gas. Additional uses for this methane include selling it directly to an industrial user, and refining it to pipeline-grade, high-BTU methane and conduction to the natural gas pipeline. Depending on economic and other variables, production of natural gas may yield a greater financial return on investment than conversion to electrical energy.

ENVIRONMENTAL AND
HEALTH BENEFITS:

Reduced risk of explosion and odor nuisance caused by the migration and emission of flammable landfill gas; enhanced recovery of biosphere above the landfill; conservation of energy resources.

Project Description

Natural Power of Raleigh, North Carolina, is controlling and extracting the gaseous product of anaerobic decomposition from organic wastes buried in the Gresham Lake landfill. The gas is removed from the ground at approximately 130 degrees F. The raw gas has a methane content which varies from 32 to 62 percent, averaging approximately 42-percent methane. The variation is due to the organic source of the gas by-product and the duration of gas extraction from a given well. It has been found that gas from certain wells maintains a consistently higher methane content than the gas from other wells. This is attributed to the type of organic refuse buried in that vicinity and, in some cases, to deeper well depth. Wells that have not been pumped for a period of time provide gas with a higher methane content than those which have been pumped recently.

A 3-horsepower vacuum pump is used to collect the gas from 10 wells sunk intermittently into the landfill. The wells are constructed of 2-inch, schedule 40, PVC pipe. The pipe is modified with fifteen 25/1000-inch perforation slots which run the length of the pipe, from the bottom of the landfill to within 15 feet of the its surface. The well pipe is sunk "naked" into drill holes, and is buffered from the landfill contents by a sand

197

and gravel fill. The wells are capped with a 5-to-1 expanding clay and with a concrete seal.

The 15 feet of landfill contents not actively mined for gas serve as a buffer which contains the biogas in the landfill and prevents the invasion of oxygen into the anaerobic environment. A 3- to 5-foot surface of compacted clay caps the landfill. The edges of the landfill are especially well-sealed and are regularly inspected for cracks or signs of subsidence.

The currently mined landfill area is 4 acres, with an average depth of 30 feet (6.3 million cubic feet). The collected gas is tapped from a 4-inch diameter pipe which is angled down and away from the generator building. This design permits easy removal of the condensate; it is collected at the far end of the 4-inch pipe and is periodically drained when the facility is not in operation.

The raw gas is of sufficient quality and quantity to power two model 3306 Caterpillar diesel engines for 15 hours a day, 5 days a week. It is calculated that the gas is consumed at a rate of 2,684 cubic feet per hour. Because of the variation in gas quality, the weekly curve for kW generation runs from a high of 182 kW on Mondays to 140 kW on Fridays. Because the quality of the gas does vary, the generators must be constantly monitored to optimize their performance. The AC current generated at the site is characterized as 3-phase, 60-cycle at 480 volts.

The capital cost to install the collection system, electrical system, and generators was $125,000. The operator claims, however, that this cost could be reduced to $100,000, based on the knowledge gained by installing the Gresham Lake system. The plant currently operates at slightly less than 50 percent of maximum capacity, but is designed for expansion into additional landfill acreage (18 acres).

Electricity generated by the engines is metered and fed directly into the Carolina Power and Light Company (CP&L) transmission lines. Operation times are set so that maximum energy generation takes place during peak demand hours when a higher rate is received per kWh. Rates are generally either "fixed", with a higher price per kWh, or "variable". CP&L offers a 15-year contract for fixed rates. In late 1984, the variable rate dropped from 3.7 cents per kWh to 2.8 cents per kWh. This decline has reduced the profitability of the operation and extended its payback period. System maintenance is minimal. Maintenance and other down times are planned for off-peak hours.

PROCESS DIMENSIONS
(1984 dollars)

==

WASTES GENERATED	Reduced venting of methane
PRODUCTIVITY	Additional revenue from landfill
PRODUCT QUALITY	Not applicable
DOWN TIME	Information not provided
FACILITIES	$125,000 one-time cost
EQUIPMENT	Included in facilities cost
RAW MATERIALS	No cost
WATER	None
ENERGY	$25,716 yearly revenue (1985 dollars)
WASTE DISPOSAL	None
POLLUTION CONTROL	None
PERSONNEL/MAINTENANCE	Approximately $6000 cost per year
NET BENEFITS	Low-cost, renewable energy generation and reduced risk of landfill explosions from methane accumulation

==

Case Study 39

TYPE OF INDUSTRY: Municipal Wastewater Treatment

NAME OF FIRM: Hickory/Newton/Conover Complex
(Burmeister, Wright & Associates, consultants,
John E.R. Burmeister 704/525-6051)

LOCATION: Fairgrove Wastewater Treatment Plant
Catawba County, North Carolina

SIC: 4952

```
****************************************************************
*                                                              *
*                       MODIFICATION                           *
*                                                              *
*           MUNICIPAL SEWAGE SLUDGE COMPOSTING                  *
*                        (PLANNED)                             *
*                                                              *
****************************************************************
```

Abstract

A composting system for municipal waste sludge is planned
for the Hickory/Newton/Conover Complex in Catawba County. The
facility is in the final design phase and is scheduled for
construction in 1986. When operational, the system will be the
first mechanical enclosed composting system in North Carolina.
The reuse of waste sludge as a composting material is expected to
save the county as much as $827,000 in annual disposal costs.

ANNUAL COST SAVINGS: Not determined.

PAYBACK PERIOD: Not determined.

TIME TO IMPLEMENT: 1-1/2 years for design and
construction.

TRANSFERABILITY: Since municipal solid waste can no
longer be automatically landfilled,
many cities are looking to land
disposal, composting, and incineration
as possible alternatives. Because
incineration is capital-intensive, it
is often not a feasible option. It
also precludes any reuse value for the

200

sludge biomass. There are several
types of composting operations. The
mechanical system to be used by Catawba
County is more expensive than some
other methods, but it is less labor-
intensive. A static pile system is
cheaper to set up (see City of
Morganton), but requires more labor and
involves more worker exposure.

ENVIRONMENTAL AND
HEALTH BENEFITS: Conservation of land and material
                               resources, improved physical properties
                               of the soil receiving compost. There
                               is a risk that unprotected piles might
                               become a play area for children or
                               pets. Because of variation in the
                               metal content of composted sludge, it
                               is not approved for use in vegetable
                               gardens or in areas where direct-line
                               food crops are to be grown.

Project Description

     The Hickory/Newton/Conover Complex in Catawba County was
seeking an alternative to their landfill disposal of municipal
wastewater sludge.   In 1981, they performed a preliminary
investigation to evaluate incineration, composting, and land
application. Based on this study, composting was selected as the
most efficient and cost-effective method.

     The firm of Burmeister, Wright and Associates was retained
to design a composting operation for the region. Several
composting designs were evaluated with respect to capital cost,
operation and maintenance costs, revenue from composted material,
and environmental impacts. Based on their evaluation, the firm
recommended an enclosed composting system. When the facility is
completed it will be the first mechanical enclosed system in
North Carolina.

     Five commercial mechanical systems were evaluated by the
consultants. The most important criteria for evaluation were the
mechanical complexity of each system and the energy costs
required for operation. The "mechanical enclosed system" will
include reactors in which the actual composting process will take
place; a carbonaceous storage tank; air blowers and scrubbers; a
conveyor and solids handling system; and, associated buildings
and structures. Sludges, from several treatment plants and
septage systems in the county, will be transported to a regional

facility for dewatering.   Sludge will be delivered to the regional plant in tankers. Four tankers will suffice initially, although it is anticipated they will have to be replaced within 10 years and an additional tanker purchased. Septage will be hauled by individual pumpers.

The proposed composting facility will be designed to meet the needs of the participating communities over the next 20 years, and will have the ultimate capacity to process 20 tons of dry sludge per day.   At start-up, the system will process approximately half this amount.

Capital costs for the entire system are estimated to be $6,135,000. This amount will be offset by federal grants of as much as 75 percent of the eligible costs. Local savings will be proportionate to the amount of the grant and to the actual quantity and price of the compost sold.

One way to calculate the savings would be to assume that 75-percent funding is received for the capital costs, and that, on the average, 75 percent of the compost is sold at $20 per ton over the life of the facility.   The net annual cost of the compost system would then be $250,000 per year versus $1,077,000 using the present disposal practices, or an average net savings of $827,000 per year.

The project has been placed on the federal funding list and is scheduled to begin construction in October 1985, the start of the 1986 federal fiscal year. The North Carolina Department of Natural Resources and Community Development will reimburse local governments for the funds used in preparing the project plans and specifications. Plan preparation has begun and the plans were submitted to the Department in the second half of 1985.

PROCESS DIMENSIONS

==

WASTES GENERATED	Will eliminate 10 tons of dry waste sludge per day
PRODUCTIVITY	Information not provided
PRODUCT QUALITY	Information not provided
DOWN TIME	Information not provided
FACILITIES	$6,135,000 estimated
EQUIPMENT	One-time cost including equipment
RAW MATERIALS	Bulking materials will be required
WATER	No change anticipated
ENERGY	Information not provided
WASTE DISPOSAL	Estimated savings of $800,000 per year
POLLUTION CONTROL	No change anticipated
PERSONNEL/MAINTENANCE	Information not provided
NET BENEFITS	Conservation of land and material resources, minimization of waste disposal costs

==

Case Study 40

TYPE OF INDUSTRY: Municipal Waste Treatment

NAME OF FIRM: City of Morganton
                    (Carl D. Hennessee 704/437-8863)

LOCATION: P.O. Drawer 430
                    Morganton, North Carolina 28655

SIC: 4952

```
*  ----------------------------------------------------------  *
*                                                              *
*                     MODIFICATION                             *
*                                                              *
*            COMPOSTING MUNICIPAL SEWAGE SLUDGE                *
*                                                              *
*  ----------------------------------------------------------  *
```

Abstract

     The Catawba River Pollution Control Facility in Morganton,
North Carolina, has implemented the first full-scale composting
operation in the state. A mixture of municipal waste sludge and
bark is composted using a static pile system designed by city
personnel. They have eliminated the need for landfilling the
waste sludge and have created a system to reuse an otherwise
undesirable product. The compost is marketed to the public with
such success that the city is unable to keep up with consumer
demand. Although they do not make money on the operation, the
city's sludge management costs are below those estimated for
the other systems they investigated, including land application
and incineration. In response to the success of the Morganton
project, the state is developing a permit system for composting
operations to ensure safe levels of heavy metals and the absence
of pathogens.

ANNUAL COST SAVINGS: Not applicable.

PAYBACK PERIOD: Not applicable.

TIME TO IMPLEMENT: 5 years, including extensive
                        preliminary study.

TRANSFERABILITY: Morganton was the first municipality
                        in North Carolina to develop sewage

204

sludge composting. As such, they have done a lot of experimentation that other municipalities can now build on. Their decision to try a composting operation came only after careful evaluation and cost comparison with land application systems. Depending on geographic, climatic, transportation, and land availability conditions, both composting and land application might be feasible alternatives. Although the city designed its own system, there are also several commercial systems on the market. The Morganton composting operation uses a static pile system. There are also enclosed mechanical systems which are more expensive, but less labor-intensive.

ENVIRONMENTAL AND HEALTH BENEFITS:

Conservation of land and material resources, and improved physical properties of soil receiving compost. In heavy, textured clay soils, the added organic matter increases the soil's permeability to water and air, and minimizes surfacewater runoff. Addition of sludge compost to clay soils has also been shown to reduce compaction (i.e., lower the bulk density) and to increase root development of plants. Composted sludge may become a nuisance if mismanaged. There is a risk that unprotected piles might become a play area for children or pets. Due to variations in the metal content of composted sludge, it is not approved for use in vegetable gardens or in areas where direct-line food crops are to be grown.

Project Description

Morganton's Catawba River Pollution Control Plant began operation in January 1975. The facility is a pure-oxygen-activated sludge plant, with a capacity of 8 million gallons per day. It generates approximately 1,700 tons of dry sludge per year. Solids handling is accomplished by two 80-foot Eimco clarifiers, aerobic digestion, and dewatering with two Sharples

P-5000 centrifuges. As was the case with many of the facilities constructed about this time, the only plans for managing the sludge consisted of burying it in a sanitary landfill.

Fear of groundwater pollution and the aesthetic problems of disposing of this semi-solid material prompted the North Carolina Division of Health Services and Burke County to refuse to accept sludge for disposal in the county-operated landfill. These decisions were made during construction of the Catawba River Plant. Therefore, upon completion of the plant, the city was forced to construct temporary lagoons until a permanent solution could be developed.

In February 1976, with the help of Mr. Bob Carlisle and Mr. Leslie Miller of the North Carolina Agricultural Extension Service, direct land application to two state-approved tracts was begun. Sludge was applied at various rates to silage cornland belonging to Broughton Hospital and to grazingland owned by a local farmer. A three-year program was developed for each tract, along with data concerning yields from various application rates, and analyses of the sludge and soil. The cornland experiment is still being performed; however, the grassland experiment was terminated after two years because of the distance from the plant.

During the experiments with direct land application, personnel were also experimenting with a pilot-scale composting operation. Early experiments consisted of hand-mixing sawdust, bark, soil, and leaves with sludge, followed by a final mixing with a coal stoker. A conveyor system with a mixing screw was later used to proportion sludge and sawdust. By this time, information was also becoming available about the EPA's pilot projects for composting in Bangor, Maine, and at the U.S. Department of Agricultural Research Center in Beltsville, Maryland.

In 1978, the City of Morganton authorized a sludge management study to review the existing facilities and to determine the most feasible method of sludge disposal. The principal alternatives were land application, composting, and incineration. The study concluded that the handling costs for each dry ton of sludge would be $49, $59 and $141 for land application, composting, and incineration, respectively.

During the next several months the city's staff investigated the land application solution. A large number of sites of sufficient size were identified; however, topography, excessive hauling distances, close proximity to streams, and opposition by property owners quickly eliminated all but three of the sites. These problems, along with increasing fuel costs, the lack of final EPA guidelines for sludge disposal, and the unavailability of farmland prompted a closer look at the composting process. City personnel then visited the composting project at the U.S. Department of Agricultural Research Station in Beltsville, Maryland. This facility was set up in 1972 as a co-operative experiment between the Research Center, Maryland Environmental Services, and the Environmental Protection Agency.

After a careful review of the data, it soon became evident that while land application theoretically appeared to have a slight cost advantage, the composting process offered a more dependable solution and a more marketable, manageable final product. After review with the city council, Morganton decided to implement the composting operation and to maintain land application as a backup solution. The capital cost for composting (estimated at $104,000) was lower than the cost for land application ($442,000). The composting operation could also be implemented much sooner, thereby eliminating the need for construction of additional lagoons.

After an in-plant study to plan the logistics and space requirements of a composting operation, a site of slightly less than one acre was chosen at the rear of the existing wastewater treatment plant. Site preparation which consisted of blacktopping and of installing a run-off drain system, was begun in February 1980.

The Morganton composting process consists of mixing a centrifuged sludge cake (14- to 15-percent solids) with a bark bulking agent. The latter provides free air space and ensures a continuous supply of oxygen throughout the pile. Perforated steel pipes are laid on the asphalt pad and covered with a layer of bark. The mixture is piled on this layer, to a height of 7 to 8 feet, and covered with a blanket of finished compost for insulation and odor control. Each pile represents two to three days of sludge accumulation. The mix is presently a ratio of 9 yards of bark to 4 of sludge. The bark used is a mixture of new and previously composted bark. A Sweco separator with a one-inch by one-half-inch mesh screen is used to recycle wood chips and bark from a local de-barking operation.

Air is drawn through the perforated pipes by blowers, and is exhausted through a pile of finished compost for deodorization. The operator measures the temperature and oxygen level in the compost pile with probes, and adjusts the blowers to maintain a 5-percent oxygen content and minimum temperature of 140 degrees F for four days. The temperature normally increases to 160 degrees after four days and remains constant for the remainder of the 21-day composting period. Some drying also results from the passage of air through the pile. Blowers are normally operated for 4 minutes out of each 20-minute cycle. At the end of the composting period, the pile is taken down and cured for an additional 28 days. Prior to screening, a thin layer of compost is spread on asphalt for further drying. The material is screened through either a one-half-inch or one-inch screen. The coarse bark is reused for composting and the final material is finished compost. The system processes about 4.5 dry tons of compost per day.

Numerous studies have shown that, in a properly controlled composting process, practically all the common pathogenic organisms normally associated with sewage sludge are destroyed or reduced to insignificant levels. Salmonellae, the common food-poisoning organism, can regrow to a limited extent. Although many organisms of the salmonellae group possess the biological ability to repopulate compost after the thermal sterilization phase, the organism does not compete well with other microorganisms present. According to studies conducted by the EPA and the U.S. Department of Agriculture, "no significant hazard should be associated with their presence in compost so long as the compost does not come in contact with food."

The City of Morganton has prepared specific guidelines for employees working with the composting system in order to prevent any adverse health effects. These guidelines state that:

1) A medical history shall be compiled for all prospective employees.

2) Persons hypersensitive to allergens, such as dust, fungus, animal hair, etc., or persons with respiratory conditions, such as asthma, emphysema or tuberculosis, shall not be permitted to work on the site.

3) Protective clothing will be provided to all on-site employees and shall be changed before leaving the site.

4) Respirators shall be provided and their use encouraged during dry weather periods. Front-end loaders must be equipped with air-conditioned cabs.

Each consumer also receives a brochure describing how to use the compost.

The compost presently produced at Morganton contains 1.67-percent nitrogen, 0.38-percent phosphorous and 0.018-percent potassium. Although this is not equivalent to commercial fertilizer, the material serves as a good soil conditioner, mulch, and potting medium. The finished product is environmentally safe, does not have an unpleasant odor, and is an excellent soil amendment. The city believes that by using good marketing tools and maintaining effective quality control, a sufficient demand will be created to establish and maintain a competitive price for the product. They anticipate that it will become a significant revenue source capable of offsetting a large portion of the operation's costs.

The city is presently selling their composted sludge, called Morganite, to the general public for $4.50 per cubic yard. This price is equivalent to approximately $20.00 per ton; production cost is $53.75 per ton, based on annual operating and maintenance costs alone. The selling cost was purposely kept low in order to acquaint the public with the product, and in an effort to eliminate the stigma attached to anything relating to sewage sludge. As a result of media coverage, the number of repeat customers, from homeowners to commercial farmers, is growing. Acceptance of the product has been extremely favorable.

Sludge compost can be advantageous when used in potting mixes, on lawns, and as a mulch. It can also be used as a substitute for topsoil in land reclamation and public works projects, turfgrass production, nursery production of trees and ornamental plants, on golf courses and cemeteries, for revegetation of disturbed lands (e.g., from surface mining), and for landscaping parks and public buildings.

Applying sludge compost at fertilizer rates (i.e., the nitrogen requirement of the crop) to marginal soils can produce significantly higher yields than commercial fertilizers applied alone and at the same nitrogen level. Higher yields are attributed to an improvement in the physical properties of the soil. Several commercial container nurserymen are now purchasing screened compost as a potting medium. This was promoted by a horticultural experiment by Mr. Dick Bir and Mr. Vinnery Boniminio of the North Carolina Agricultural Extension Service. At a site on the grounds of the Morganton plant, azalea, japanese holly, juniper, rhododendron, and other varieties of one-year landscape specimens were planted in a medium consisting of 100-percent screened compost. A temporary irrigation system was installed, and the plants are checked periodically to determine growth. To date, the growth and general health of the plants has been comparable to the results produced by commercial potting mixes, and at only a fraction of the cost.

Additional experiments which are either under way or being considered in conjunction with the composting program include bagging the product, trying to decrease the drying time required for screening, capturing the hot air from the process blowers as a form of heat energy, and using the compost as a combustible fuel.

There is no official permitting system for compost, but the North Carolina Department of Natural Resources is working with the City of Morganton to develop a facility permit to certify the sludge. The permit application will require extensive analyses for heavy metals and pathogens.

PROCESS DIMENSIONS

===

WASTES GENERATED	Eliminated approximately 1,700 dry tons of sludge per year
PRODUCTIVITY	No change
PRODUCT QUALITY	No change
DOWN TIME	No change
FACILITIES	$47,615 one-time cost
EQUIPMENT	$94,535 one-time cost
RAW MATERIALS	$58,650 cost per year
WATER	Information not provided
ENERGY	$75.00 cost per year
WASTE DISPOSAL	$55,434 saved per year
POLLUTION CONTROL	Not applicable
PERSONNEL/MAINTENANCE	$63,411 cost per year
NET BENEFITS	Minimization of cost for environmentally sound management of sewage sludge

===

A detailed 1984-85 budget for the composting operation is as follows:

Labor (including fringe benefits)	$ 45,724
Vehicles	
Fuel	8,431
Maintenance	5,330
Screen and Blowers	
Electricity	75
Maintenance	1,840
Plastic pipe and couplings	936
Bark (assumes 2.2:1 bark/sludge ratio and 50% return)	58,650
Miscellaneous (advertising, brochure, lab analysis)	1,150
TOTAL OPERATION AND MAINTENANCE COSTS	$122,136

Case Study 41

TYPE OF INDUSTRY:	Wastewater and Sewage Treatment
NAME OF FIRM:	Chemical and Coordinate Bonding and Adsorbtion Demonstration Plant San Diego Regional Water Reclamation Agency (Robert L. Smith III, 612/228-1490)
LOCATION:	Santee, California
SIC:	4952

```
*                                                                    *
*                         MODIFICATION                               *
*                                                                    *
*   CHEMICAL REMOVAL AND FIXATION OF SEWAGE SOLIDS TO PRODUCE        *
*      MARKETABLE BUILDING MATERIAL AS AN IMPROVEMENT OVER           *
*    CONVENTIONAL BIOLOGICAL TREATMENT AND DISPOSAL OF SLUDGE        *
*                                                                    *
```

Abstract

The chemical and coordinate bond and adsorbtion system (CCBA) is a process for treating a municipal waste (raw sewage). The equipment design and operation is based on chemical processes rather then traditional biological decomposition and stabilization. Common clay, alum, and polyacrylic acid are added to the sewage influent. The chemical interaction of these compounds with the dissolved and solid sewage waste is effective in removing of 91 percent of the suspended solids, and 88 percent of the biological oxygen demand from the waste water. Remaining organics in the plant effluent are water-soluble and easily biodegradable. Sludge, resulting from the cleaning of the influent wastewater, is mixed with sufficient clay and then baked in a kiln to produce a lightweight aggregate suitable for reuse in the building industry. Additional processing of the treated wastewater permits significant removal of nutrient wastes. The processing facility is significantly less expensive to construct than conventional biological sewage treatment facilities, and is not sensitive to toxic constituents in the wastewater. An additional benefit accruing from this process is the ability to safely treat hazardous inorganic and organic chemicals in the sewage sludge by incinerating them at elevated temperatures; the residues are subsequently entrained in a ceramic matrix.

ANNUAL COST SAVINGS: Variable, not quantified.

PAYBACK PERIOD: Not applicable.

TIME TO IMPLEMENT: Information not given.

TRANSFERABILITY: Widely applicable to many industries
generating wastes best managed by
incineration and matrix entrainment;
transferable to all centralized sewage
treatment systems of 300 gallons per day or
more; can be modified for smaller systems.

ENVIRONMENTAL AND
HEALTH BENEFITS: Reduced emission of volatile organic
chemical vapor from treatment facility;
increased efficiency in resource use by
incorporating sewage sludge into building
material; insult to environmental systems
from conventional sludge disposal
eliminated, health risks from the same also
eliminated.

Project Description

The San Diego Water Reclamation Agency has constructed and is
testing a 100,000 gallon-per-day chemical and coordinate bonding
and adsorbtion (CCBA) sewage treatment plant. The CCBA process was
devloped by George C. Harrison, Ph.D., while working as a senior
chemist at the 3M company. The process is truly innovative
insofar as toxic constituents in the sewage cannot interfere with
operation and are bound into a ceramic matrix suitable for
building material; water effluent from the process meets federal
secondary treatment standards without biological treatment. July
8, 1983 was the first complete day of pilot operations. Since
that date, studies have demonstrated the feasibility of the
process.

The treatment process begins by screening the influent sewage
through 1/8-inch wedgewire for degritting; residence time is only
a few seconds. Screened materials are delivered to the tub, or
mogul, mixer which will also ultimately receive the settled
sludge. A clay slurry is added to the influent sewage water, and
the mixture is delivered to the "tea cup" where gravity separation
removes grit and sand from the mixture. Residence time is one
minute. The process uses an amount of clay, on the order of 500
parts per million (ppm), which is twice as heavy as the suspended
dry solids. Usually two to three tons of clay are used per
million gallons of water treated.

Alum is added at 4 to 100 ppm, along with 1 ppm of polymer polyacryic acid (the glue base), to the tea-cup effluent in the sheet flow mixer. Residence time is several seconds. The specially designed "sheet mixer" was developed to optimize the mixing rate within the confines of the flocculant's ability to sustain agitation without disassociation. The sheet mixer is, in addition, far more energy efficient than conventional, high-horsepower mixing "churners". No back-mixing occurs, and all turbulent energy dissipation is hydraulic.

Effluent from the tea cup is delivered to the decay gradient flow flocculator which optimally reduces the energy dissipation rate relative to the growth of the floc. This process minimizes the disruption in flocculant growth and eliminates mechanical shear, permitting the floc's unchecked growth. Residence time in this process is four minutes.

Careful testing and modeling of the flocculation process determined that the orthokinetic flocculant aggregation rate was linearly dependent on the viscous velocity gradient, proportional to the square root of the total specific-energy dissipation rate. Breakup of the floc, in contrast, is linearly related to the turbulent specific-energy dissipation rate. Theoretically, in systems without turbulent energy dissipation, floc breakup should not occur.

The floc is settled in a clarifier to form a thick sludge (10- to 12-percent solids). The colloidal solids are 98-percent removed in the flocculation procedure. The clean water is decanted off the top. The clarifier retention time is 45 minutes. Total suspended solids are reduced from the average influent quality of 232 to 20.5 milligrams (mg) per liter (91 percent), and the biochemical oxygen demand (BOD) is reduced from an average of 188 to 23.5 mg per liter (88 percent). Effluent leaving the clarifier has a good clarity of 8-14 NTUs. From the total data it is projected that a 95-percent reduction in suspended solids and a 90-percent reduction in BOD could be achieved with a clarifier retention time of under 60 minutes.

The organic materials that are not removed by the CCBA process are extremely and preferentially water-soluble. They are relatively simple chemicals of natural- or petroleum-based synthesis. To be preferentially water-soluble they must be alcohols, glycols, glycerine, sugars, amines, or salts of simple organic acids. Sewage rarely exceeds 20 ppm of these soluble organics, and they are, in any event, the most readily biodegradable wastes in the sewage stream.

With the addition of adequate alum, 95-percent phosphate removal can be achieved. Additional processing of the decanted water, by filtering it through material such as rice hull char, is effective in the problematic removal of nitrogen, achieving an efficiency of 43 percent. The use of buoyant macrosphere filter media on the surface of the clarifer can further reduce suspended solids to 1 ppm.

The sludge mixture is pumped from the bottom of the clarifier into a filter, and pressed into a cake. A cake of 37.3-percent solids is easily achieved, with a 45-percent solids measurement achievable when a vibratory unit is added. The sludge end-product is then mixed with more clay (screened to 1/8 inch) in the mogul mixer until a putty-like composition suitable for extrusion is achieved. The sludge-clay solid is extruded through a die and cut into pellet particles. The pellet is tumbled to round the shape, and is then heated in a kiln at 2,000 degrees F to form what is known as "low weight aggregate" (LWA).

Making LWA from clay is a well-known technology throughout the United States and Europe. The process incorporates a "bloating agent" into the clay, such as peat or coal, which causes the pelletized clay mixture to expand in the kiln. The machinery and equipment necessary to prepare the clay, mix the additives, pelletize the mixture, and form the pellets into aggregate have been well developed by industry over the past decades.

The CCBA process has produced LWA that is comparable to the highest quality, commercially produced LWA. Six of the 25 prototype LWAs were used to make test concrete sections. The aggregate has a loose bulk density of 800 to 2000 pounds per cubic yard. Compressive strengths in excess of 5,000 pounds per square inch were achieved, with wet concrete densities ranging between 98 and 115 pounds per cubic foot. The ceramic incorporates the ash of the combusted organic waste and any heavy metals present in the sewage. Water and the strongest acid conditions found in nature will not leach any compounds from the ceramic pellet.

Waste heat from furnaces in the region can be used to fire the "sewage pellets". As another possibility, furnaces that combust this refuse-derived fuel could manufacture steam and electricity, thus linking two elements to form a comprehensive waste treatment plan.

CCBA plants are smaller than sewage plants using microbiological means to purify wastewater. They are also simpler to design and less costly to build. The capital cost of a 10 million gallon-per-day CCBA facility is estimated at $0.80 per gallon of treatment capacity ($8 million). Discounting the sale of products from the process, the costs of capital, operation, and maintenance are estimated to run $484 to $538 per million gallons processed. This is approximately half the capital cost alone of a conventional, activated sludge, secondary treatment facility. Typically, conventional biological wastewater treatment facilities are priced at about $3.50 per gallon of capacity.

Phase 1 of the CCBA demonstration project has, in the 14-month period preceding the first summary report, scientifically documented that:

1) Raw sludge from wastewater can be reclaimed as a valuable lightweight aggregate.

2) The capital cost is approximately 25 percent that of an activated sludge, secondary treatment facility.

3) The operation and management costs of the plant may be greatly reduced by revenue from the sale of the recovered resources.

4) Environmental impacts are minimized.

One process design consideration which requires more attention is ammonia reduction. The ammonia content of CCBA effluent water is nearly as high as that of effluents from microbiological sewage treatment plants. CCBA reduces ammonia nitrogen by only 15- to 25-percent. By adding autotrophic biological filter beds which use anaerobic microbes, nearly all ammonia can be removed. A 10-ppm residual of nitrate from respiration after processing is typical; however, the nitrogen in this form provides for biologically stable water (it cannot, on its own, support aerobic microbiological growth). The nitrogen can be further reduced by filtering the effluent through a bed composed of another type of autotrophic microbiota known as denitifying bacteria. These bacteria readily use nitrate and organic materials, usually considered aerobically biorefractory, for respiration. The products of respiration are gaseous nitrogen and carbon dioxide.

Overall, test results from the facility show that it is converting raw sewage to reusable water of a quality that will more than meet secondary discharge requirements. The resulting ceramic pellets are of a high-grade LWA, with a market value of $35 to $65 per ton in the southern California area. The market, is however, international in scope. With regard to resource conservation and waste reuse, clay can be obtained from sand and aggregate mining operations which wash their products clean of the clay fraction. Such clay, once screened for uniformity, has been shown to be adequate for the pelletizing process.

An additional possibility for increased resource efficiency may be realized by placing heat-exchange pumps on the CCBA effluent in order to recover vast amounts of heat. Sewage water typically measures 65-plus degrees 365 days a year, in the temperate U.S. Skimming the top 10 degrees at a high coefficient of performance can provide sizable district heating source. Another way to coordinate a comprehensive waste management program would be to use refuse-derived fuel to stoke the kilns for LWA production.

PROCESS DIMENSIONS

WASTES GENERATED	All sewage wastes eliminated, disposal of solid residues not required
PRODUCTIVITY	Less than 5 tons of LWA per million gallons influent treated
PRODUCT QUALITY	Water effluent superior to conventional process, LWA pellets comparable to conventional products
DOWN TIME	Information not provided
FACILITIES	Cost of $0.80 per gallon of treatment capacity (10 million gallons per day)
EQUIPMENT	Estimated to be 20 percent of the cost of biological plants
RAW MATERIALS	Decreased cost, savings variable depending on the availability of clay resources
WATER	Not applicable
ENERGY	Substantial savings, not quantified
WASTE DISPOSAL	Significant savings, potential for income
POLLUTION CONTROL	Savings, not quantified
PERSONNEL/MAINTENANCE	Savings, not quantified

219

NET BENEFITS   Overall savings on plant construction,
                                operation, and maintenance of up to 50
                                75 percent of the cost of conventional
                                processing, significant conservation of
                                resources, near total elimination of
                                environmental and health effects
                                associated with sewage treatment

==

Case Study 42

TYPE OF INDUSTRY: Lube Oil Packaging and Warehouse

NAME OF FIRM: Chevron USA Inc.
                (Steve Raatz 502/566-5607)

LOCATION: Louisville, Kentucky

SIC: 5171

```
*****************************************************************
*                                                               *
*                        MODIFICATIONS                          *
*                                                               *
*    1)   HIGH-PRESSURE WATER SYSTEM REPLACED CAUSTIC DRUM       *
*         RINSING SYSTEM                                         *
*                                                               *
*    2)   WATER-BASED PAINT AND WATER CURTAIN FILTER FOR         *
*         OVERSPRAY REPLACED SOLVENT-BASED PAINT IN PAPER        *
*         FILTERING SYSTEM                                       *
*                                                               *
*****************************************************************
```

Abstract

Chevron USA, Inc., maintains a petroleum product warehouse in Louisville, Kentucky. A shipping-drum reconditioning facility, where used drums are cleaned and repainted for reuse, is operated on-site. Prior cleaning practices generated a complex hazardous waste stream, composed of caustic, oil, water, and paint which cost Chevron $50,000 per year to manage. A high-pressure, hot-water cleaning system was installed to replace the caustic cleaning system. The new system is a closed-loop wastewater process which reuses the filtered cleaning water and ensures the recycling of oil cleaned from the drums. A substitution of oil-based paints with a water-base process, using state-of-the-art overspray filtration equipment, conserves paint and eliminates the generation of hazardous waste paint residue. The total savings to Chevron from these two modifications exceed $80,000 per year, after the 3-year payback period.

ANNUAL COST SAVINGS: $115,500

PAYBACK PERIOD: 3 years

TIME TO IMPLEMENT:	3 months for planning, 3 months for ordering and installation
TRANSFERABILITY:	These techniques are widely transferable to a broad range of SIC codes. The drum-cleaning strategy could be equally well applied to equipment requiring similar standards in other industries which handle oil. Similarly, the water-based spray paint system is already receiving wide use in a number of SICs that had previously used solvent-based paints with the overspray system of paper filtering.
ENVIRONMENTAL AND HEALTH BENEFITS:	Safety, fire, and health problems created by the old, drum reconditioning system have been eliminated; resources are conserved and toxic wastes are no longer disposed of in the environment.

Project Description

The Chevron USA lube oil warehouse and packaging plant in Louisville, Kentucky, has an on-site, 55-gallon-drum cleaning and painting facility. Prior to the modifications of the drum reconditioning facility, drums were washed with a caustic liquid and rinsed with water. This resulted in the generation of 500 gallons per day of a caustic, oil, paint, and water mixture. The caustic rinsate mix did not meet the local sewer district's wastewater discharge limitations and had to be disposed of as a hazardous waste, costing the company $50,000 a year.

Similarly, paper filters which collected the overspray from the solvent-based paint spraybooth also posed a disposal problem. Furthermore, the filters did not provide adequate air emission controls, and additional measures to reduce solvent emissions from the paint spraybooths were required. Employee exposure to the noise, caustic vapors, and solvent paint-base were described in a company report as unacceptable. Over a 3-month period the company evaluated a variety of prospective solutions to alleviate these conditions, including upgrading the existing system, and subcontracting drum reconditioning to an outside vendor.

Economic considerations, however, favored the installation of a state-of-the-art drum washing, reconditioning, and painting system. The new system uses only high-pressure hot water to wash the drums. The oil is reclaimed and the water is reheated and reused. A small volume of wastewater is discharged to the local sewer district and is within the wastewater discharge limitations. The reclaimed oil is sold to a recycler, netting Chevron $12,000 per year. The organic solvent-based paint was replaced with a water-based paint; a water-wall filtration system captures the overspray. The filtrate residue is non-hazardous. Efficiency has been greatly enhanced by the new system; down time has been virtually eliminated; plant capacity was increased by 700 drums per day; and, savings of $25,000 per year in personnel costs, and of $6,800 per year in maintenance and other equipment costs have been realized. Additionally, caustic and other chemical material costs have been reduced by $13,400 per year, with a similar savings of $5,300 in yearly paint costs.

PROCESS DIMENSIONS

==

WASTES GENERATED	Eliminated 100 percent of previously generated hazardous wastes
PRODUCTIVITY	Increased plant capacity by more than 200 percent per day
PRODUCT QUALITY	No change
DOWN TIME	Reduced by nearly 100 percent (4 to 5 hours per week)
FACILITIES	$390,000 one-time cost (includes equipment)
EQUIPMENT	Included above
RAW MATERIALS	$18,700 saved per year
WATER	No change
ENERGY	$15,000 saved per year
WASTE DISPOSAL	$50,000 saved per year
POLLUTION CONTROL	Not quantified. Reduced cost of maintaining breathable air in the plant due to the removal of toxic oil-based paint fumes, and reduced wastewater treatment costs.
PERSONNEL/MAINTENANCE	$31,800 saved per year

NET BENEFITS   Cost savings, waste toxicity reduced,
                                improved conditions in the workplace,
                                and eliminated production and disposal
                                of hazardous waste

==

Case Study 43

TYPE OF INDUSTRY: Film Developing

NAME OF FIRM: Minnesota Mining and Manufacturing (3M)
                     (Electronics Products Division,
                      Mike Koeningsberger 612/778-4523)

LOCATION: Columbia, Missouri

SIC: 7374

```
*——————————————————————————————————————————————————————*
*                                                        *
*                    MODIFICATION                        *
*                                                        *
*   WASTEWATER TREATMENT SYSTEM ALTERED FROM SINGLE-PASS *
*     SYSTEM TO CLOSED-LOOP WITH RECOVERY OF SOLVENT     *
*                                                        *
*——————————————————————————————————————————————————————*
```

Abstract

A film developing unit at 3M's Electronics Products Division was discharging wastewater contaminated with 1,1,1-trichloroethane. In order to recycle the solvent and to continue in compliance with process water discharge regulations, 3M installed a decanter system that provides gravity separation of the solvent from the water. This modification saved the company $12,000 in its first year of operation by reducing the amount of new solvent and makeup water required for the developing unit.

ANNUAL COST SAVINGS: $12,000

PAYBACK PERIOD: 3 months

TIME TO IMPLEMENT: Information not provided.

TRANSFERABILITY: Widely applicable to industries
                         producing waste streams containing
                         organic solvents, especially 1,1,1-
                         trichloroethane, given sufficient
                         economies of scale.

ENVIRONMENTAL AND
HEALTH BENEFITS: Reduced risks of surface- and
                         groundwater contamination by solvent;

226

conservation of water and solvents;
reduced chance of solvent adversely
affecting municipal wastewater
treatment process; reduced chances of
non-worker contact with solvent.

Project Description

In order to ensure continued compliance of plant discharges
with water effluent regulations and to recover a valuable
solvent, 3M modified their film developing unit in Columbia,
Missouri.   The unit was discharging a mixture of water and
1,1,1-trichloroethane into the municipal sewage system.   3M found
a better way of dealing with the solvent in the water by
installing a decanter system which provides gravity separation of
the solvent in a closed-loop with the developing unit.   This
system provides a short hold-time during which the heavier
solvent settles out of the rinse water.   The decanter system cost
$4,000, including installation, and has saved the company $12,000
in its first year by reducing the amount of new solvent and
makeup water required for the developing unit.

PROCESS DIMENSIONS

===

WASTES GENERATED	1,1,1-trichloroethane removed from process water
PRODUCTIVITY	No change
PRODUCT QUALITY	No change
DOWN TIME	Decreased
FACILITIES	No additional costs
EQUIPMENT	$4,000 one-time cost
RAW MATERIALS	Cost savings, reduced demand
WATER	Cost savings, reduced demand
ENERGY	Information not provided
WASTE DISPOSAL	No added cost
POLLUTION CONTROL	No added cost
PERSONNEL/MAINTENANCE	Minimal
NET BENEFITS	Increased efficiency, maintenance of quality of process water sent to municipal sewage treatment plant, cost savings

===

Case Study 44

TYPE OF INDUSTRY: Photographic Processing

NAME OF FIRM: PCA International, Inc.
                    (Fred Monroe 704/847-8011)

LOCATION: Matthews, North Carolina

SIC: 7395

```
*-------------------------------------------------------------*
*                                                             *
*                      MODIFICATIONS                          *
*                                                             *
*       1)    ELECTROLYTIC SILVER RECOVERY                    *
*                                                             *
*       2)    DEVELOPER REGENERATION BY ION EXCHANGE          *
*                                                             *
*-------------------------------------------------------------*
```

Abstract

PCA International generates several waste streams during their photographic processing of color portraits. They have developed a number of modifications to reduce their waste volume. Electrolytic recovery is used to remove silver from film and paper processing solutions. The recovered silver is 96-percent pure and the recovered fixer can be replenished and reused. Ion exchange columns are used to remove the inorganic chemical, bromide, from the color developer solution. This system has allowed 85-percent reuse of the color developer. Spent bleach solution is restandardized by chemical addition and is recycled, allowing 90-percent reuse.

ANNUAL COST SAVINGS: Variable due to fluctuating silver
                        prices.

PAYBACK PERIOD: 1 year (electrolytic recovery)
                        Less than 1 year (ion exchange)

TIME TO IMPLEMENT: Information not provided.

TRANSFERABILITY: Optimum silver recovery is desirable
                        because the metal has a high market
                        value and, if left unrecovered in a
                        waste stream, is a pollutant. The

229

cost of in-house recovery equipment is high, however, and a small firm may elect to send waste solutions out-of-house for metal recovery, or to form a cooperative to purchase and share one set of equipment. (See Case Study 45.) Out-of-house recovery requires more bookkeeping and more transportation risks. The water and solution regeneration/reuse modifications undertaken by PCA pioneered these activities for the industry nationwide. Equipment manufacturers can assist small firms in evaluating the potential payback for similar installations on a smaller scale.

ENVIRONMENTAL AND
HEALTH BENEFITS:

Reduced pollutant load to municipal sewage treatment system; conservation of water and materials; reduced threat of silver compounds entering surface- and groundwater.

Project Description

PCA International is a color portrait processing company. Their waste stream is characterized by high biological and chemical oxygen-demand concentrations. When their Matthews facility was first constructed in 1972, a goal of PCA's management was to minimize the pollution load they contributed to the small, municipal wastewater treatment facility. The result has been the development of several projects which have significantly reduced waste generation and yielded annual savings for the company.

An electrolytic process is used to recover silver from color negative film fixer solutions and from the paper fixer solutions. Spent fixer solutions from both processes are mixed and passed into a continuous-flow silver recovery unit. A large cathode rotates in the solution, and a low voltage current (3 to 4 volts) is applied. The silver collects on the cathode. Once a week the cathodes are removed and the silver knocked off in chunks. The desilvered solution is aerated, returned to full strength, and pumped back to the processing units. The system allows approximately 96-percent reuse of the fixer solution; raw material costs have been reduced by about $25,000 per year. The silver chip obtained is about 96-percent pure, and approximately 2,200 troy ounces of silver are recovered each week.

Overflow waste fixer solution is collected in a tank and subjected to an electrolytic silver recovery batch process. Effluent is then passed through 5-gallon canisters containing steel wool for a final polishing. In the presence of thiosulfate complex in the fixer, the iron in the steel wool becomes soluble, and the silver from the complex precipitates out as metallic silver. The canisters can usually recover about 80 to 85 percent of the silver in solution; the silver concentration of the original fixer solution is reduced from 1,500 grams per liter to 0.02 grams per liter after passing through both recovery systems. About 24 canisters are used each year, at a cost of $35 each. The spent canister systems are sent to a refiner for silver recovery, and each yields approximately 150 troy ounces of silver.

Canister systems are also used to remove silver from one of the paper processing rinses. The rinse, called the "slow wash", is the first wash after the fixer solution, and is performed at a lower flow than the other wash operations. Slow-wash water is collected and passed through two cartridges in series, yielding 6- to 8-percent silver recovery. The wash water then exits to the municipal sewer system.

A second modification undertaken by PCA was the regeneration of color developer by ion exchange. As photographic materials are developed, they release bromide into the developer solution. This release can inhibit the development process, and requires the replacement of the developer solution when the bromide levels get too high. PCA has been able to regenerate spent color paper developer by removing the excess bromide with ion exchange columns. Developer solution, containing 0.8 to 1.0 grams of bromide per liter is passed through a sand filter, then through an absorption resin to remove contaminants, and finally through two sets of ion exchange columns for bromide removal. The effluent contains 0.07 grams of bromide per liter. The reclaimed developer is readjusted to the required concentration before it is returned to the processing tanks. The ion exchange columns are regenerated on-site using an 8-percent sodium chloride solution. This system has allowed 84-percent reuse of the color developer solution.

PCA uses approximately 8,500 gallons of color paper developer each week. The ability to regenerate the spent solution has reduced costs from $1.40 per gallon to $0.57 per gallon, saving about $360,000 per year.

231

Spent color developer from film processing is collected and stored for a short time in case it is needed for reuse. Currently, it is not routinely recycled.

Spent bleach solution is restandardized by the addition of necessary chemicals and reused. Ninety percent of the bleach solution can be recycled by using this rejuvenation process, saving over $2.50 per gallon in raw material costs.

PCA uses countercurrent flow for their rinse water in order to minimize the number of washes required and to enhance product quality. They also have a well-developed housekeeping program to minimize spillage and loss of chemicals. In the rinse operations, rubber "squeegees", similar to automotive windshield wipers, are used to remove water and chemicals adhering to the film. This reduces the dragout from tank to tank and maintains solution integrity as long as possible. The "squeegees" are wrung dry regularly to maintain their efficiency.

Other housekeeping features include careful labeling of chemicals; dedication of pumps to certain chemicals to reduce cleaning operation wastes; direct pumping rather than manual transfer of bulk chemicals to reduce risk of spillage; use of drum lifts to minimize spillage and accidents; and, use of larger drums for chemical storage to reduce spillage and cost.

Over the past ten years, PCA has also experimented with several other modifications to reduce waste generation. These have included:

1) Using reverse osmosis (RO) units to recycle rinse waters from film and paper processing. This was abandoned because of reduced product quality.

2) Silver in the concentrated wastes collected on the RO system was recovered by electrolysis preceded by ion exchange. Wastes were passed through a strong anion exchange column which increased the silver concentration from 0.5 grams per liter to 5 grams per liter. Electrolytic silver recovery was then performed on the solution, resulting in 97-percent pure silver chip.

3) Concentrating the wastes collected on the RO system by using spray evaporators. A film spray evaporator system received the wastewater at a rate of 15,000 gallons per day. The concentration of soilds was increased from 5 percent to 30 percent on a single pass through the system. Further cycles through the system yielded a 60- to 70-percent solids cake which could then be sold for silver recovery.

PROCESS DIMENSIONS

===

WASTES GENERATED	Removed 115,000 troy ounces of silver per year from wastes. Eliminated approximately 1,700 gallons of color developing waste per day. Eliminated approximately 19 gallons spent fixer waste per day. Eliminated approximately 1,200 gallons of spent bleach waste per day.
PRODUCTIVITY	No change
PRODUCT QUALITY	No change
DOWN TIME	No change
FACILITIES	$24,000 (electrolytic recovery) Complete information not provided
EQUIPMENT	$96,000 (electrolytic recovery) Complete information not provided
RAW MATERIALS	$360,000 saved per year (developer) $25,000 saved per year (fixer) $780,000 saved per year (bleach)
WATER	No change
ENERGY	$200 cost per year (electrolytic recovery)

WASTE DISPOSAL	Information not provided
POLLUTION CONTROL	Information not provided
PERSONNEL/MAINTENANCE	No change
NET BENEFITS	Increased revenue from silver recovery, cost savings from solution reuse

===

Case Study 45

TYPE OF INDUSTRY: Photographic Fixer and Film Processing for Silver Recovery

NAME OF FIRM: American Fotokemi, Inc.
(Bernt Hermansson 312/228-0755, telex: 206834)

LOCATION: Elk Grove, Illinois

SIC: 7399

```
*  ------------------------------------------------------  *
*  |                  MODIFICATIONS                     |  *
*  |                                                    |  *
*  |  1) CENTRALIZATION OF PROCESSEING                  |  *
*  |                                                    |  *
*  |  2) IMPROVED MONITORING EFFICIENCY                 |  *
*  |                                                    |  *
*  |  3) IMPROVED CHEMICAL PROCESSES                    |  *
*  ------------------------------------------------------  *
```

Abstract

American Fotokemi, Inc., provides a central processing plant for silver recovery from films and spent fixer solutions used in film processing. Conventional approaches to silver recovery in the film processing industry have located canister and electrolytic recovery equipment at each individual film processing site. This decentralized approach results in a less than optimal recovery of silver due to sub-ideal process monitoring and to the operating economics of small, independent recovery units. Small processors who cannot economically justify recovery units lose all their silver. American Fotokemi has developed chemical treatment processes that increase silver recovery to nearly 100 percent; an increase of 10- to 15-percent from conventional decentralized technology. Furthermore, other toxic metals -- which would otherwise be discharged as pollutants from the small independent, silver recovery operations -- are also recovered by American Fotokemi treatment processes. The high efficiency of this centralized recovery system has increased the dollar/silver return for everyone who has converted to it from in-house canister and/or electroplating recovery.

The statistics cited below are for a client of American Fotokemi, who is located in Chicago.

ANNUAL COST SAVINGS: $6,000, plus prior expenses for private equipment, maintenance, services, etc.

PAYBACK PERIOD: Immediate.

TIME TO IMPLEMENT: Information not provided.

TRANSFERABILITY: Available to film processors within the market area of American Fotokemi; or, interested parties can license the process anywhere in the United States.

ENVIRONMENTAL AND
HEALTH BENEFITS: Reduces loading of heavy metals wastes to the environment by 15 to 100 percent, while also conserving valuable resources. Eliminates potential for toxic metal poisoning by removing metals from effluent discharge. Eliminates worker hazard associated with some silver recovery methods by using safer processing technology.

Project Description

American Fotokemi (A.F.) selected the Chicago area to test how receptive the film developing market would be to its new, centralized silver recovery operation. A.F. offers its clients the distinct advantage of not having to invest in any equipment. Instead, A.F. provides, at no charge, holding tanks for storing used fixer solution, and collects the solution at regular intervals for processing at their facility.

At the A.F. processing facility, the solution is treated in electroplating units, 315 gallons in each tank per day. In this step approximately 95 percent of the silver is recovered. To extract the remaining silver, a chemical precipitation step is performed. First the pH is brought to between 7 and 8, from the acid condition of 4, which is common for spent fixer solution. Then another chemical is added to convert all metals to their metallic form. The metals then precipitate as a sludge. After removal, the sludge is dried and refined, and the remaining solution is neutralized. Another closely related opportunity for silver recovery exists in extracting the silver from used photographic film and photographic paper. Several processes are presently used, predominantly burning and cyanide removal, although an expensive and dangerous enzymatic method also exists. These processes have various negative attributes which limit their usefulness. American Fotokemi and a Scandinavian company have developed a new chemical method for recovering silver from film and paper. A very fast-working solution removes all the silver by destroying the gelatin binding layer on the film base. Because it works so quickly, paper can also be washed effectively. The silver remains in complex form, so it can be easily filtered out of solution.

The potential for increased silver recovery on a nationwide basis is excellent and the profit to be realized is staggering. Hospitals are estimated to use approximately 41 percent of the 100 million pounds of photographic film and paper sold in the U.S. Of the 7 million troy ounces of silver represented, conservative estimates are that at least .10 percent is lost in poor refining (cyanide or burning). It is further estimated that 16,320,000 gallons of fixer are used to develop the film and paper, representing another 6,528,000 troy ounces of silver. Assuming each film developing lab practices conventional silver recovery, a conservative average of 20 percent is still rinsed down the drain. If the recovery processes described here were in use across the country, 2,005,600 troy ounces of silver could be saved from discharge into sewage systems and into our environment, from hospitals alone. If all photographic developing labs were included, the figure would be closer to 4,890,000 troy ounces of silver, worth an estimated $48.9 million per year. Because the reported contractual arrangements between American Fotokemi and its clients return 70 percent of the silver value -- in collected fixer, film, and paper -- to the client, the photographic developing industry could expect to boost its earnings, as a whole, by $33 million.

Equally important is the health threat represented by the more than 2 million troy ounces (1.6 tons) of silver which are presently discharged to our sewage treatment plants and receiving waters each year. Silver is classified for toxicological purposes as a group 1 heavy metal, along with mercury and copper. It is among the most toxic of heavy metals, though little is known about its effects, except in the field of marine biology. Many sea animal groups, economically important as links in our food-chain, are known to be sensitive to silver. Furthermore, as another indication of its toxicity, silver filters are increasingly used, as a replacement for chlorine, to control bacterial growth in swimming pools.

PROCESS DIMENSIONS

==

WASTES GENERATED	Reduced
PRODUCTIVITY	No change
PRODUCT QUALITY	No change
DOWN TIME	No change
FACILITIES	No investment required
EQUIPMENT	No investment required
RAW MATERIALS	No change
WATER	Information not provided
ENERGY	Information not provided
WASTE DISPOSAL	Information not provided
POLLUTION CONTROL	Greatly enhanced, costs not applicable
PERSONNEL/MAINTENANCE	Not applicable
NET BENEFITS	Enhanced conservation profitablity

==

Taken from a paper by Bernt Hermansson, Environmentally Safe Recovery of Silver from Photowaste Material, presented at the conference: "Pollution To Profit: Reducing Industrial Waste in Illinois," April 16-17, 1984.

Case Study 46

TYPE OF INDUSTRY: Automotive Remanufacturing

NAME OF FIRM: Vulcan Automotive Equipment Ltd.
                    (James Meikle 604/876-3384)

LOCATION: Vancouver, British Columbia, Canada

SIC: 7699

```
*  ----------------------------------------------------  *
*                                                        *
*                      MODIFICATION                      *
*                                                        *
*    REPLACEMENT OF CAUSTIC CLEANING PROCESS WITH ABRASIVE *
*                  ALUMINUM SHOT SYSTEM                  *
*                                                        *
*  ----------------------------------------------------  *
```

Abstract

Vulcan Automotive Equipment Ltd. remanufactures used automotive engines. The cleaning process was modified by replacing the inorganic caustic cleanser with a high-velocity "aluminum shot" system. This modification in equipment design resulted in substantial savings to the company due to decreased costs for raw materials, labor, and to avoided waste management costs. The new aluminum shot system increases productivity and improves the final product. All by-products of the new process are recycled.

ANNUAL COST SAVINGS: $40,000 per year

PAYBACK PERIOD: Approximately 2 years

TIME TO IMPLEMENT: 6 months

TRANSFERABILITY: This process is used to clean cast
                        metal parts which have accumulated
                        rust, carbon, and oily grime. Given
                        adequate volume to justify the capital
                        investments, this technology is
                        applicable to any shop engaging in the
                        reprocessing of engines and other
                        similar cast-metal components.

ENVIRONMENTAL AND
HEALTH BENEFITS:
            Greatly enhanced work environment, with exposure to caustic substances eliminated. Environmental health is improved because hazardous wastes which were previously buried are no longer generated.

Project Description

Vulcan Automotive Equipment Ltd. remanufactures and restores used automotive engines. As an initial step in refurbishing components, they are cleaned of caked oil, grease, carbon, and other deposits of grime. The process originally used to clean the metal parts involved the wet application of caustic soda, followed by scrubbing and rinsing the parts. The workplace was characterized as "mucky", and presented potentially hazardous exposure to caustic liquid and vapor.

A caustic sludge, generated as a by-product of this process, was stored on-site in tanks. The contents of the tanks were emptied two or three times a year, at a cost of $12,000 to $18,000 per year. Water requirements for rinsing were 48,000 gallons per year. Six 55-gallon drums of caustic soda were required yearly, at a total cost of $5,400.

This process was replaced by a two-step system. The first step involves baking the metal parts to remove volatile oils and grease. In the second step, the parts are sprayed with a high-velocity stream of aluminum shot which thoroughly cleans them of grime and rust. In this step, the parts also acquire an aluminum coating which is rust-resistant and which imparts an appearance of "brand-new cast metal".

All aspects of the new, aluminum shot system are superior to the old caustic soda process. The work crew needed to clean and inspect the parts previously numbered three. With the advent of the new system only two men are required. One inspector is now able to do the work of two, with the other man now 10-percent more productive. Because the new system removes rust much more thoroughly than the old process, parts are easier to inspect for cracks.

Canadian law and the lack of an adequate burial facility had previously forced Vulcan to ship the caustic hazardous waste to a U.S. facility for burial. The aluminum shot system produces no hazardous wastes. The aluminum dust by-product can be recycled. This saves $12,000 to $18,000 per year in caustic sludge waste management costs. A cost comparison shows the savings realized by the new system: raw material costs for caustic soda were $5,400 per year; the aluminum shot costs only $1,500 per year. Additionally, the cost of water was eliminated, as the aluminum shot system does not require it. The overall costs of operation have been reduced from $17,000 per year to $15,000 per year, excluding labor. Labor costs were reduced by $25,000.

The new system has improved the overall appearance of the business and has substantially reduced health hazards to the workers. Pollution control measures are now necessary only for scrubbing emissions from the baking oven. However, analysis of baking oven residues for lead content has identified an area that should be investigated.

PROCESS DIMENSIONS

WASTES GENERATED	Eliminated
PRODUCTIVITY	Enhanced by 45 percent
PRODUCT QUALITY	Improved
DOWN TIME	Not applicable
FACILITIES	$5,000 one-time cost
EQUIPMENT	$70,000 one-time cost
RAW MATERIALS	Approximately $4,000 saved per year in change from caustic soda to aluminum
WATER	Need eliminated (48,000 gallons per year)
ENERGY	No change
WASTE DISPOSAL	Hazardous waste generation eliminated, $12,000 to $18,000 saved per year
POLLUTION CONTROL	Information not provided
PERSONNEL/MAINTENANCE	$25,000 saved per year
NET BENEFITS:	Conservation, elimination of waste disposal costs, improved worker conditions, decreased operation and labor costs, threat of environmental toxics eliminated, improved product quality

CHAPTER 5

Abstracts Cross-Referencing Case Studies by Materials

In this chapter the case studies' abstracts are arranged to highlight the classes of material subject to waste reduction and resource conservation. This arrangement facilitates the examination of pollution prevention strategies appropriate for the particular material which is of interest to the reader. The pollution prevention process used to effect the waste minimization is similarly cross-indexed in Chapter 6.

The materials are indexed according to the materials which are underlined in the case studies' abstracts. These key words are underlined in the case studies' abstracts in Chapter 4 along with the process used. In this chapter only the process is underlined to give the appropriate context for each abstract.

The categories of material identified with respect to the case studies presently included in this compendium are:

AGRICULTURAL/FOOD PROCESSING WASTES

FLY ASH AND BOTTOM ASH

LOW-LEVEL RADIOACTIVE WASTES

METALS AND OTHER INORGANIC CHEMICALS

MUNICIPAL WASTES

PROCESS WATER

RUBBER AND PLASTICS

SOLVENTS, OILS, AND OTHER ORGANIC CHEMICALS

AGRICULTURAL/FOOD PROCESSING WASTES

CASE STUDY 1

The Choplin Family Farm raises swine and various field crops. In 1982, the Choplin farm saved $8,000 in fertilizer costs by spraying swine wastes on crops with a standard irrigation gun. They also reduced organic material losses by 1,670 percent, total nitrogen losses by 490 percent, and total phosphorus losses by 525 percent through the integrated use of conservation tillage, parallel terracing, grassed waterways, and cover crops. The reuse of these agricultural wastes and the management initiative to incorporate conservation practices in the farm operation resulted in corn crop yields that were 300 percent above the county average.

CASE STUDY 4

Dunbar Foods processes canned peppers and sweet potatoes. They have developed a spray irrigation system for land application of their food processing wastes. Process water containing vegetable wastes is reused to irrigate and fertilize grassland owned by the company.

CASE STUDY 5

Food processing wastes generated during beer brewing operations at Miller Brewing Company are treated and made available for land application. The by-product is used for liming and as a nitrogen supplement. The reuse of this previously landfilled material has eliminated disposal costs for Miller, provided a virtually cost-free substitute for standard agricultural liming material, and preserved valuable landfill capacity.

FLY ASH AND BOTTOM ASH

CASE STUDY 36

Carolina Power and Light Company (CP&L) has been selling waste fly ash and bottom ash for reuse since 1973. The company does not sell ash directly to users but has contracted with another firm, Monier Resources, Inc., to oversee the marketing of the ash. Prior to this arrangement, all of the ash produced was sluiced to on-site ash ponds. The company has now reduced its disposal costs and prevented the burial of usable material.

CASE STUDY 37

Since the late 1960s, Duke Power Company has been marketing waste fly ash and bottom ash from their coal-fired power plants for reuse. The ash has been used for concrete production, asphalt filler, and plastics filler. The company is exploring other uses for the ash, particularly the possibility of using it as structural backfill or to stabilize roadbase.

LOW-LEVEL RADIOACTIVE WASTES

CASE STUDY 37

Duke Power Company has instituted a comprehensive program to reduce the volume of their low-level radioactive wastes. The program includes housekeeping improvements; the purchase of equipment to allow cleaning rather than disposal of contaminated tools and equipment; conversion from an evaporator-based, non-recyclable waste processing system to one based on ion exchange; redesign of existing evaporator equipment so that it can process recyclable waste; and, review and revision of filtration equipment operation to improve performance and to minimize replacement and disposal costs.

METALS AND OTHER INORGANIC CHEMICALS

CASE STUDY 2

In order to reduce the amount of water that had to be treated before discharge and to recover the soluble molybdenum from the tailings stream, Climax Molybdenum Company studied the natural drainage system surrounding the tailing impoundment area. They then installed interceptor canals to channel upstream runoff waters from around the tailing system into natural streams. The tailings runoff water is passed through a countercurrent ion exchange unit in order to remove the molybdenum, which is recovered and sold. Other heavy metal separation from process water is accomplished by the Swift Lectro Clear process, which removes over 90 percent of the iron, manganese, zinc, and copper from the tailings stream by a coagulative electrolytic process. Cyanide, used as a copper depressant in the regrind plant, is 96-percent removed.

CASE STUDY 11

Allied Corporation manufactures chemicals. A $4.5 million investment in recycling equipment is paying off at a rate of 1 million dollars a year for Allied Corporation. Allied's Metropolis, Illinois, plant recovers 8,000 tons of calcium fluoride annually. The orgrganic chemical is then used as a raw material in anhydrous hydrofluoric acid production at another facility. Since 1982, over 1,000 cubic yards of hazardous calcium fluoride wastes have been recycled monthly, saving about $300,000 a year in disposal and storage costs alone. The process also recovers about 1,000 tons of lime annually.

CASE STUDY 17

IAC uses materials substitution to reduce their raw material costs in the manufacture of trace-element compounds. Suitable inorganic chemical wastes are received from firms that would otherwise pay to landfill them. Three examples are cited describing the utility of this arrangement to both IAC and to the donor of the waste material.

247

CASE STUDY 20

Texasgulf manufactures phosphate products. In designing their new defluorinated phosphate plant to meet EPA regulations pertaining to the release of fluorides into the air and water, they decided to implement a closed-loop process water system to remove inorganic fluorides from the discharge stream. Other alternatives considered included releasing the partially treated wastewater to Saltville's municipal sewage system, or discharging it into the nearby Holton river. Benefits of a closed-loop rather than a single-pass system to recycle process water, include:

1) A savings of approximately 280,000 gallons of water per day;

2) A combined annual savings in water purchases and treatment of $930,000 per year;

3) Compliance with effluent guidelines without a NPDES permit, since there is no longer any industrial discharge.

CASE STUDY 22

Florida Steel Corporation sells metal dust collected from furnace fumes to a zinc smelter. This practice has reduced their disposal costs by almost half, and allows reuse of the metal rather than disposal.

CASE STUDY 24

The Stanadyne Company undertook a comprehensive and systematic review of their electroplating processes in order to reduce or eliminate waste generation. The result of their efforts has been a broad spectrum of activities which have saved the firm money and minimized pollution. A sulfuric/peroxide materials substitution for chromic acid reduced sludge production in the parts cleaning operation. The company also found they could operate their copper- and chrome-plating bath equipment with a lower concentration of cyanide, saving both raw material and sludge handling costs. Other equipment design and operation changes, which reduced dragout volumes and rack recoating costs, also improved plating efficiency. Tanks were installed, before the rinsing operation, to recycle valuable metals back to the plating baths. A change in filtration equipment operation for the plating bath drastically reduced solution losses, and resulted in a 93-percent cost reduction in filter media. A filter press operation for sludge dewatering minimized the cost of sludge transport and disposal. A new wastewater filtration system has improved the quality of treated process water. Housekeeping improvements led to leak elimination and to a preinspection program which prevents the plating of defective parts.

CASE STUDY 25

Elkhart Products, Inc., manufactures pipe fittings. Under the guidance of Lancy International, they have modified their bright-dipping and passivating-line wastewater treatment system. It was replaced with a newly developed pickling process that uses sulfuric acid and a proprietary, stabilized hydrogen peroxide. By modifying the equipment operation, Elkhart was able to remove a cyanide dip and chromic acid dip from the process line. They also recover metal with an electrolytic cell which recovers copper at the rate of 1,000 to 1,500 pounds per month.

CASE STUDY 26

Pioneer was using a single-pass system to treat their plating wastewaters. Because the system had no large end-tank to correct minor upsets in the system, concentrations of metals and other inorganic chemicals in the effluent regularly exceeded water quality regulations. The effluents were killing many species of plants and animals in the receiving marsh. Pioneer solved this problem by removing the old single-pass process water treatment system and installing a closed-loop batch system manufactured by Zerpol. As a result of the new equipment design, the company now uses only 500 gallons of water per day (makeup due to evaporation), instead of the previous 12,000 gallons per day, and has experienced a 20-percent reduction in the amount of sludge produced. Plans are being made to sell the sludge to a company for use as a filler in the manufacture of concrete products.

CASE STUDY 27

Deere and Company reduced its hazardous waste by 80 percent in volume and by over 99 percent in level of risk through two key actions:

1) Implementation of a comprehensive, hazardous waste management initiative at each manufacturing unit.

2) Construction of an on-site facility for treating hazardous liquid waste. The facility reclaims waste organic oil compounds and metals from the process water, and produces a non-hazardous sludge. The <u>recovered</u> oil is sold to an oil recycling firm or is <u>reused</u> for machining processes.

CASE STUDY 30

With stricter regulations concerning copper effluents soon to be enforced, GTE, aided by Lancy International, decided to modify the treatment of process water from their copper-etching rinse stream by installing an <u>electrolytic recovery</u> cell. The electrolytic cell is part of a closed-loop system that follows the etching operations on the Electroless copper-plating line. The cell was installed downstream from the rinse tank, where it recovers copper that would have ended up as a landfilled sludge. After the copper ions are removed, the rinse water is recirculated back to the rinse tank. The copper is periodically removed from the cell's cathode and sold as No. 1 copper scrap. In its first year of operation, this system produced $2,000 from the sale of recycled copper scrap, and saved $4,000 in avoided waste disposal costs.

CASE STUDY 31

Data General Corporation generates several hazardous waste streams during the manufacture of printed circuit boards. Most of these wastes are characterized by high heavy-metal concentrations. In 1981, the company launched a vigorous program to reduce their reliance on hazardous waste landfills. A <u>management initiative</u> to develop markets for their untreated wastes resulted in a net profit of $9,000 per year from combined by-product sales revenue and avoided waste managment costs. Treatment of other wastes using new process chemistry, including changes in <u>equipment design or operation</u>, rendered additional material suitable for sale for metal recovery. Data General's program to produce salable wastes and to market those wastes has reduced the hazardous material they landfill by over 400 tons per year, and has yielded a cost savings of $180,000 per year.

CASE STUDY 32

3M's microelectronics plant makes flexible electronic circuits from copper sheeting. Before use, the metal is cleaned. Previously, this was accomplished by spraying it with ammonium persulfate, phosophoric acid, and sulfuric acid. This process produced hazardous wastes that were expensive to manage properly. An alternative equipment design was introduced to replace the chemical cleaning process. The new process features a machine with rotating brushes that mechanically clean the metal with pumice. The fine, abrasive pumice leaves a sludge that is not hazardous and that can be placed in a conventional sanitary landfill. This new method has reduced 3M's production of hazardous waste by 40,000 pounds a year. They also saved $15,000 in the system's first year of operation due to reduced raw material, disposal, and labor costs.

CASE STUDY 33

Digital Equipment Corporation's new, printed wiring board manufacturing operation integrates several resource conservation and recovery methods in its process water treatment system. The system, designed by Lancy International, includes metal recovery by electrolytic cell and by evaporation. The electrolytic reaction recovers 92 percent of the dissolved copper from dragout rinses, and the atmospheric evaporator recovers 95 percent of the chromic acid dragout. The chromic acid is recycled into the etch process line. The company also use a lime slurry by-product from a local acetylene gas manufacturer as an acid neutralizer.

CASE STUDY 34

Modine prepares brass parts for the manufacture of radiators by dipping them in a series of chemical solutions. The hydrogen peroxide/sulfuric acid bright dip was causing them several problems, including the buildup of copper sulfate pentahydrate crystals that were then deposited in a hazardous waste landfill. There was, in addition, a significant dragout of copper ions into the subsequent rinse tanks. Rather than incur the initial and long-term landfilling costs for the hazardous wastes, Modine elected to eliminate the formation of the crystals by continuously recirculating the hydrogen peroxide/sulfuric acid bright dip solution through an ion exchange column. The column selectively removes the metal. The purified dip solution is recycled to the bath. Once the ion exchange resin becomes saturated, the copper ions are exchanged again in the regeneration process, and are subjected to an electrolytic reaction to remove them from the solution. The plated copper is sold as No. 1 copper scrap.

CASE STUDY 41

The chemical and coordinate bond and adsorbtion system (CCBA) is a process for treating a municipal waste (raw sewage). The equipment design and operation is based on chemical processes rather then traditional biological decomposition and stabilization. Common clay, alum, and polyacrylic acid are added to the sewage influent. The chemical interaction of these compounds with the dissolved and solid sewage waste is effective in removing of 91 percent of the suspended solids, and 88 percent of the biological oxygen demand from the waste water. Remaining organics in the plant effluent are water-soluble and easily biodegradable. Sludge, resulting from the cleaning of the influent wastewater, is mixed with sufficient clay and then baked in a kiln to produce a lightweight aggregate suitable for reuse in the building industry. Additional processing of the treated wastewater permits significant removal of nutrient wastes. The processing facility is significantly less expensive to construct than conventional biological sewage treatment facilities, and is not sensitive to toxic constituents in the wastewater. An additional benefit accruing from this process is the ability to safely treat hazardous inorganic and organic chemicals in the sewage sludge by incinerating them at elevated temperatures; the residues are subsequently entrained in a ceramic matrix.

CASE STUDY 44

PCA International generates several hazardous waste streams during their color portrait processing operations. Electrolytic recovery removes silver from fixing and bleach-fix solutions. The recovered silver is 96-percent pure, and the fixer can be replenished and reused. Ion exchange columns are used to remove the inorganic chemical, bromide, from color developer solution; this system has allowed 99-percent reuse of color developer.

CASE STUDY 45

American Fotokemi, Inc., established a central processing plant to recover silver from films and spent fixer solutions used in film processing. Conventional approaches to silver recovery in the film processing industry have located canister and electrolytic recovery equipment at each individual film-processing site. This decentralized approach results in less than optimal recovery of silver due to sub-ideal process monitoring, and to the operating economics of small, independent recovery units. There is a 100-percent loss of silver by small processors who cannot economically justify recovery units. American Fotokemi has developed chemical treatment processes that increase silver recovery to nearly 100 percent -- an increase of 10 to 15 percent over that achieved by conventional, decentralized technology. Furthermore, other toxic metals, which would otherwise be discharged as pollutants by the small, independent silver-recovery operations, are also recovered. The high efficiency of this centralized recovery method has increased the dollar/silver return for everyone who has converted to it from in-house canister and/or electroplating recovery.

CASE STUDY 46

Vulcan Automotive Equipment Ltd. remanufactures used automotive engines. The cleaning process was modified by substituting the inorganic caustic cleanser with a high-velocity "aluminum shot" system. This modification in equipment design resulted in substantial savings to the company, due to decreased costs of raw materials, labor, and to avoided waste management costs. The new, aluminum shot system increases productivity and improves the final product. All by-products of the new process are recycled.

MUNICIPAL WASTES

CASE STUDY 38

Natural Power is a small generator of electricity operating from the Gresham Lake Landfill in Raleigh, North Carolina. They have implemented a landfill gas recovery system for converting biogas to electric power; the electricity is then sold to the local utility. Natural Power recovers combustible gas generated from the anaerobic decomposition of municipal waste and other organic material buried at the landfill. The gas is collected, filtered and fed directly into two diesel generators modified to use a methane fuel source. Electrical energy generated at the landfill is metered and fed into the local electric utility's transmission lines.

CASE STUDY 39

A composting system for municipal waste sludge is planned for the Hickory/Newton/Conover Complex in Catawba County, North Carolina. The facility is in the final design phase and is scheduled for construction in 1986. When operational, the system will be the first mechanical enclosed composting system in North Carolina. Reusing waste sludge as composting material is expected to save the county as much as $827,000 in disposal costs annually.

CASE STUDY 40

The Catawba River Pollution Control Facility in Morganton, North Carolina, has implemented the first full-scale composting operation in the state. A mixture of municipal waste sludge and bark is composted using a static pile system designed by city personnel. They have eliminated the need for landfilling the waste sludge, and have created a system to reuse an otherwise undesirable product. The compost is marketed to the public and the city is unable to keep up with consumer demand. Although they do not make money on the operation, their sludge disposal costs are below those estimated for other systems investigated, including land application and incineration. In response to the success of the Morganton project, the state is developing a permit system for composting operations which will insure safe levels of heavy metals and the absence of pathogens.

CASE STUDY 41

The chemical and coordinate bond and adsorbtion system (CCBA) is a process for treating a municipal waste (raw sewage). The equipment design and operation is based on chemical processes rather then traditional biological decomposition and stabilization. Common clay, alum, and polyacrylic acid are added to the sewage influent. The chemical interaction of these compounds with the dissolved and solid sewage waste is effective in removing of 91 percent of the suspended solids, and 88 percent of the biological oxygen demand from the waste water. Remaining organics in the plant effluent are water-soluble and easily biodegradable. Sludge, resulting from the cleaning of the influent wastewater, is mixed with sufficient clay and then baked in a kiln to produce a lightweight aggregate suitable for reuse in the building industry. Additional processing of the treated wastewater permits significant removal of nutrient wastes. The processing facility is significantly less expensive to construct than conventional biological sewage treatment facilities, and is not sensitive to toxic constituents in the wastewater. An additional benefit accruing from this process is the ability to safely treat hazardous inorganic and organic chemicals in the sewage sludge by incinerating them at elevated temperatures; the residues are subsequently entrained in a ceramic matrix.

PROCESS WATER

CASE STUDY 2

In order to reduce the amount of water that had to be treated before discharge and to recover the soluble molybdenum from the tailings stream, Climax Molybdenum Company studied the natural drainage system surrounding the tailing impoundment area. They then installed interceptor canals to channel upstream runoff waters from around the tailing system into natural streams. The tailings runoff water is passed through a countercurrent ion exchange unit in order to remove the molybdenum, which is recovered and sold. Other heavy-metal separation from process water is accomplished by the Swift Lectro Clear process, which removes over 90 percent of the iron, manganese, zinc, and copper from the tailings stream by a coagulative electrolytic process. Cyanide, used as a copper depressant in the regrind plant, is 96-percent removed.

CASE STUDY 7

Two major hazardous waste streams, generated during United Globe's furniture manufacturing are incinerated to produce steam and heat. Solids from the ragwash and spraybooth process water wastes are thickened using dissolved air flotation. The resulting sludge is mixed with woodchips, dried, and fed to the primary combustion chamber of a two-stage incineration system. Semi-solid and liquid finishing wastes, including high-BTU solvents, are fed to the secondary combustion chamber. The steam produced is used for space heating, ragwashing/drying operations, and the sludge dryer. Use of the incinerator has reduced United Globe's waste volume by 95 percent, and the ash produced is not hazardous. Discharge loading to the municipal sewer system has decreased dramatically, as have hazardous waste transport and disposal costs.

CASE STUDY 20

Texasgulf manufactures phosphate products. In designing their new, defluorinated phosphate plant to meet EPA regulations pertaining to the release of fluorides into the air and water, they decided to implement a closed-loop process water system to remove inorganic fluorides from the discharge stream. Other alternatives considered included releasing the partially treated wastewater to Saltville's municipal sewage system, or discharging it into the nearby Holton river. Benefits of a closed-loop, rather than a single-pass system to recycle process water include:

1) A savings of approximately 280,000 gallons of water per day;

2) A combined annual savings in water purchases and treatment of $930,000 per year;

3) Compliance with all effluent guidelines without a NPDES permit, since there is no longer any industrial discharge.

CASE STUDY 23

Due to planned plant expansion and new governmental regulations, Waupaca Foundry modified the equipment design of their process water treatment system from a single-pass system to two interconnected closed-loops. Through selective separation of the wastewater streams, Lancy International, the consulting firm, was able to design the closed-loop process so that only 25 percent of the plant's total process water had to be treated. This approach minimizes chemical usage in the wastewater treatment, keeps pumping costs down, and reduces water usage.

CASE STUDY 26

Pioneer was using a single-pass system to treat their plating wastewaters. Because the system had no large end-tank to correct minor upsets in the system, concentrations of metals and other inorganic chemicals in the effluent regularly exceeded water quality regulations. The effluents were killing many species of plants and animals in the receiving marsh. Pioneer solved this problem by removing the old, single-pass process water treatment system and installing a closed-loop batch system manufactured by Zerpol. As a result of the new equipment design, the company now uses only 500 gallons of water per day (makeup due to evaporation) rather than the previous 12,000 gallons per day, and has experienced a 20-percent reduction in the amount of sludge produced. Plans are being made to sell the sludge to a company for use as a filler in the manufacture of concrete products.

CASE STUDY 27

Deere and Company reduced its hazardous waste by 80 percent in volume and over 99 percent in level of risk through two key actions:

1) Implementation of a comprehensive hazardous waste management initiative at each manufacturing unit.

2) Construction of an on-site facility for treating hazardous liquid waste. The facility reclaims waste organic oil compounds and metals from process water, and produces a non-hazardous sludge. The recovered oil is sold to an oil recycling firm or is reused for machining processes.

CASE STUDY 30

With stricter regulations concerning copper effluents soon to be enforced, GTE, aided by Lancy International, decided to modify the treatment of process water from their copper-etching rinse stream by installing an electrolytic recovery cell. The electrolytic cell is part of a closed-loop system that follows the etching operations on the Electroless copper-plating line. The cell was installed downstream from the rinse tank, where it recovers copper that would have ended up as a landfilled sludge. After the copper ions are removed, the rinse water is recirculated back to the rinse tank. The copper is periodically removed from the cell's cathode and sold as No. 1 copper scrap. In its first year of operation, this system produced $2,000 from the sale of recycled copper scrap, and saved $4,000 in avoided waste disposal costs.

CASE STUDY 31

Data General Corporation manufactures printed circuit board for computers. Process water, containing high concentrations of heavy metals, is passed through a filter-based water purificatio system, and is then disposed of by application to forestland owned by the company. This was the first site for the land application of pretreated metal-plating waste in North Carolina.

CASE STUDY 33

Digital Equipment Corporation's new printed wiring board manufacturing operation integrates several resource conservation and recovery methods in its process water treatment system. The system, designed by Lancy International, includes metal recovery by electrolytic cell and by evaporation. The electrolytic reaction recovers 92 percent of the dissolved copper from dragout rinses, and the atmospheric evaporator recovers 95 percent of the chromic acid dragout. The chromic acid is recycled into the etch process line. The company also uses a lime slurry by-product from a local acetylene gas manufacturer as an acid neutralizer.

CASE STUDY 43

A film developing unit at 3M's Electronics Products Division was discharging wastewater contaminated with 1,1,1-trichloroethane. In order to recycle the solvent and to continue in compliance with process water discharge regulations, 3M installed a decanter system that provides gravity separation of the solvent from the water. This modification saved the company $12,000 in its first year of operation by reducing the amount of new solvent and makeup water required for the developing unit.

RUBBER AND PLASTICS

CASE STUDY 21

The Nebraska Small Farms Action Group, with the help of the Institute for Local Self-Reliance, studied the feasibility of recycling used tires as a secondary resource for roadpaving and other products which use rubber. The pilot-scale project used buffings from tire retreaders as representative of crumb rubber derived from the cryogenic processing of used tires. In the study, a section of road was laid using the crumb rubber in a 2-percent mixture with the conventional, asphalt surfacing material. Road life was documented to be double that of conventional asphalt, and economic studies indicated that the test material was favorably cost-competitive with concrete.

SOLVENTS, OILS, AND OTHER ORGANIC CHEMICALS

CASE STUDY 7

Two major hazardous waste streams, generated during United Globe's furniture manufacturing, are incinerated to produce steam and heat. Solids from the ragwash and spraybooth process water wastes are thickened using dissolved air flotation. The resulting sludge is mixed with woodchips, dried, and fed to the primary combustion chamber of a two-stage incineration system. Semi-solid and liquid finishing wastes, including high-BTU solvents, are fed to the secondary combustion chamber. The steam produced is used for space heating, ragwashing/drying operations, and the sludge dryer. Use of the incinerator has reduced United Globe's waste volume by 95 percent, and the ash produced is not hazardous. Discharge loading to the municipal sewer system has decreased dramatically, as have hazardous waste transport and disposal costs.

CASE STUDY 9

The Rexham Corporation facility in Greensboro, North Carolina, manufactures and prints specialized product labels. They have made several modifications in the printing operations at their various North Carolina facilities to reduce hazardous waste production. A material substitution of waterborne ink for alcohol/acetate ink in some of their applications has reduced the volume of spent solvent requiring disposal, and has lowered air emission levels. The company also installed a distillation unit to reclaim n-propyl alcohol from their waste inks. The unit reduced their hazardous waste disposal costs by 74 percent, and decreased their raw material costs by 16 percent.

CASE STUDY 10

Exxon Chemical installed floating roofs over its tanks of volatile solvent. This has prevented the loss of more than $1.5 millon worth of solvents. Concurrently, they installed conservation vents on the plant's large hydrocarbon tanks, which reduced air emissions from the tanks by between 30 and 75 percent. These two equipment design modifications have, in addition to reducing air pollution, conserved a significant amount of resources. A management initiative created a "stewardship program" to reduce fluctuations in the wastewater treatment system. Sampling stations were established to monitor concentrations of organic chemicals; as a result, operators can trace any fluctuations to their origin and quickly remedy the problem. The information gained from the monitoring is also used to charge each process with a portion of the costs of the wastewater treatment system, and to set targets for waste reduction for each process. This program resulted in a 75-percent reduction in the quantity of organic wastes entering the wastewater treatment plant.

CASE STUDY 12

Borden Chemical Company reduced the amount of organic phenol resins they discharged by 95 percent, mainly through three process modifications:

1) Phenolic resins are passed through a filtration process that removes the large resinous particles. Previously, the filtered materials were disposed of by rinsing them into the wastewater treatment system. They are now rinsed into large tanks and recycled into the process as raw material.

2) Reactor vessels were cleaned by filling them with water. This produced a large amount of phenolic resin waste that had to be processed through the wastewater treatment system. New equipment operation procedure has reduced the initial rinse to only 5 percent of the previous volume, resulting in a more concentrated solution of phenolic resins. This solution is stored and recycled into the process line as a raw material. The reactor vessels are then rinsed a second time. The waste stream from this rinse contains a low concentration of phenolic resins.

3) The Borden Company has made employees aware that small amounts of chemical waste can strongly affect overall waste generation. As a result, management and workers have identified ways of reducing wastes.

CASE STUDY 13

American Enka is a nylon yarn production and research facility. They have found it economically profitable and environmentally sound to recycle their waste isopropyl alcohol solvent in-house, rather than having it recycled by an outside firm. They purchased a used distillation unit and, with approppriate modifications, American Enka is now saving $90,000 per year. They have also been able to reuse the still bottom residues as an asphalt emulsifier in another product line.

CASE STUDY 14

Riker Laboratories was coating medicine tablets, using several organic solvents. Consequently, they had the potential for exceeding air pollution limits set by a new, California environmental regulation. A water-based solvent was developed to replace the organic solvent; different spray equipment was installed in order to handle the new solvent. This combination of resource substitution of the water-based solvent for the organic solvent, and of equipment design modification saves $15,000 per year in solvent costs and prevents the production of 24 tons of air pollution per year. Subsequently, the company determined that $180,000 in pollution control equipment for organic solvent vapors was no longer required.

CASE STUDY 15

The Coty Division of Pfizer, Inc. has developed a method of heat recovery from waste hydro-alcoholic products. The ignitable organic wastes are mixed with fuel oil and burned to produce steam and hot water. The company saves over 1,000 gallons of fuel oil a year, and the costs of hazardous waste disposal have been eliminated.

CASE STUDY 16

USS Chemicals has reduced the organic chemical emissions from their phenol manufacturing processes by a series of equipment design modifications including:

1) Adding a refrigerated condenser unit in combination with a resin adsorption system, at a cost of $5,000. This reduced emissions by 80 percent and saves 237,000 pounds of cumene per year. The cumene savings is worth $60,000 per year.

2) Adding a scrap condenser that returns the emissions to the process. This saved 400,000 pounds of cumene, worth $100,000, in its first year of operation.

3) Adding floating roofs to acetone storage tanks, and an adsorber unit for acetone recovery.

SOLVENTS, OILS, AND OTHER ORGANIC CHEMICALS, CONTINUED

Modification #2 arose from employee ideas submitted to USS Chemicals' "Suggestions for Cost Reduction" (SCORE). SCORE is a management initiative which provides plant employees with a direct financial incentive for identifying cost-cutting practices; the program rewards them with a percentage of the savings that have resulted from the implementation of their ideas. Employees have received a total of $70,000 for successful suggestions that have saved the firm $500,000 in costs.

CASE STUDY 18

ICI Americas, Inc., operates an agricultural chemical research and development lab in Goldsboro, North Carolina. As a research facility, it does not produce a large amount of waste. An Environmental Compliance Committee reviews study protocols; their primary emphasis is on early planning to minimize waste production; and secondarily, on attempting to recycle or treat the waste, whenever possible, so that none is discharged. Measures recommended by the committee, and then implemented include housekeeping improvements, and the initiation of waste solvent recovery by distillation.

CASE STUDY 19

Daly-Herring Company altered their dust collection equipment operation so that waste streams, containing organic chemicals from various production areas, are now collected separately, rather than mixed in a single baghouse. The collected materials are no longer contaminated by alternate waste streams, and each material is recycled back to the process where it originated. The firm has eliminated over $9,000 in annual disposal costs, and estimates that the recovered material is worth more than $2,000 per year.

CASE STUDY 24

The Stanadyne Company undertook a comprehensive and systematic review of their electroplating processes in order to reduce or eliminate waste generation. They also incorporated pollution prevention goals in their recent plant expansion, selecting a new degreaser which has significantly lowered solvent costs. The new unit has several features which help to lower solvent requirements and to prevent solvent losses.

SOLVENTS, OILS, AND OTHER ORGANIC CHEMICALS, CONTINUED

CASE STUDY 27

Deere and Company reduced its hazardous waste by 80 percent in volume and by over 99 percent in level of risk through two key actions:

1) Implementation of a comprehensive, hazardous waste management initiative at each manufacturing unit.

2) Construction of an on-site facility for treating hazardous liquid waste. The facility reclaims waste organic oil compounds and metals from process water, and produces a non-hazardous sludge. The recovered oil is sold to an oil recycling firm or is reused for machining processes.

CASE STUDY 28

Emerson Electric Company has made a series of modifications which have significantly decreased operating costs; increased productivity; reduced solvent, paint solids, and plating chemical costs; conserved energy; improved product quality; provided a safer working environment; and conserved raw materials. Several of the improvements involved the purchase of new equipment which allowed them to automate their electroplating process and their energy management system. Conversion from an organic solvent-based paint system to a waterborne system reduced paint solvent and paint sludge losses by 40 percent.

CASE STUDY 29

Scovill, Inc. uses 1,1,1-trichloroethane to degrease metal stampings produced in their manufacturing process. Solvent wastes are collected in 55-gallon drums and housed in a storage building designed to contain spills, if they should occur. The drums are picked up by a company that distills 1,1,1-trichloroethane from the wastes and returns it to Scovill for reuse. The recycled solvent costs about half as much as the virgin material. The company has also experimented with material substitution of water soluble synthetic cleaners; they have consequently reduced their use of organic solvents by 30 percent. A management initiative, which rewards ideas from employees for waste reduction and cost savings projects, is also part of Scovill's program for pollution prevention.

SOLVENTS, OILS, AND OTHER ORGANIC CHEMICALS, CONTINUED

CASE STUDY 35

The Rexham Corporation facility in Matthews, North Carolina, laminates and coats specialized paper, film, and foil substrates for the photography industry. A management initiative to develop a comprehensive waste management plan for the firm has resulted in several pollution prevention modifications. New housekeeping practices, which require the segregation of spent solvent by ink color and type, have resulted in almost complete reuse of liquid toluene. Toluene-laden vapors are collected by a carbon-adsorption-based, air pollution control system, and are sold for use in the coatings industry. Off-site distillation of spent solvent mixtures has allowed reuse of methyl ethyl ketone solvent for washup operations. Methyl ethyl ketone vapors are removed from the air and incinerated for heat recovery.

CASE STUDY 37

Duke Power Company has an on-going program to phase out polychlorinated biphenyl (PCB) from their electrical distribution system. Organic chemical insulating fluid contaminated with PCBs is burned as a supplemental fuel for heat recovery at one of their steam-electric generating stations. The PCBs are totally destroyed, and the company has saved about $84,000 in hazardous waste disposal costs in each of the past few years.

CASE STUDY 41

The chemical and coordinate bond and adsorbtion system (CCBA) is a process for treating a municipal waste (raw sewage). The equipment design and operation is based on chemical processes rather then traditional biological decomposition and stabilization. Common clay, alum, and polyacrylic acid are added to the sewage influent. The chemical interaction of these compounds with the dissolved and solid sewage waste is effective in removing of 91 percent of the suspended solids, and 88 percent of the biological oxygen demand from the waste water. Remaining organics in the plant effluent are water-soluble and easily biodegradable. Sludge, resulting from the cleaning of the influent wastewater, is mixed with sufficient clay and then baked in a kiln to produce a lightweight aggregate suitable for reuse in the building industry. Additional processing of the treated wastewater permits significant removal of nutrient wastes. The processing facility is significantly less expensive to construct than conventional biological sewage treatment facilities, and is not sensitive to toxic constituents in the wastewater. An additional benefit accruing from this process is the ability to safely treat hazardous inorganic and organic chemicals in the sewage sludge by incinerating them at elevated temperatures; the residues are subsequently entrained in a ceramic matrix.

CASE STUDY 42

Chevron USA, Inc., maintains a petroleum product warehouse in Louisville, Kentucky. A shipping-drum reconditioning facility, where used drums are cleaned and repainted for reuse, is operated on-site. Prior cleaning practices generated a complex hazardous waste stream, composed of caustic, oil, water, and paint which cost Chevron $50,000 per year to manage. A high-pressure, hot-water cleaning system was installed to replace the caustic cleaning system. The new system is a closed-loop wastewater process which reuses the filtered cleaning water and ensures the recycling of oil cleaned from the drums. A substitution of oil-based paints with a water-base process, using state-of-the-art overspray filtration equipment, conserves paint and eliminates the generation of hazardous waste paint residue. The total savings to Chevron from these two modifications exceed $80,000 per year, after the 3-year payback period.

CASE STUDY 43

A film developing unit at 3M's Electronics Products Division was discharging wastewater contaminated with 1,1,1-trichloroethane. In order to recycle the solvent and to continue in compliance with process water discharge regulations, 3M installed a decanter system that provides gravity separation of the solvent from the water. This modification saved the company $12,000 in its first year of operation by reducing the amount of new solvent and makeup water required for the developing unit.

CHAPTER 6
Abstracts Cross-Referencing
Case Studies by Process

In this chapter, the case studies' abstracts are arranged according to the processes used to achieve the minimization or elimination of a waste stream.

This arrangement illustrates how the processes examined in these studies can be used to reduce waste economically, in a variety of applications, and in different industries. Making the information available in this format should help industrialists to see applications for these processes that would be appropriate in their own facilities. The materials to which the processes are applied are similarly cross-referenced in Chapter 5.

The categories are representative of the process groupings which were underlined as key words along with the material in the case studies' abstracts from Chapter 4. In this chapter only the material is underlined to give the appropriate context for each abstract.

In some cases, the entire process category name is not included in the abstract, in order to enhance the readabilty. For example, reuse or recovery indicates the category: recycling, recovery, and reuse.

The process categories presently included in this compendium are:

ADSORPTION

DISTILLATION

ELECTROLYTIC REACTIONS

EQUIPMENT DESIGN AND OPERATION

FILTRATION

HOUSEKEEPING

INCINERATION/HEAT-ENERGY RECOVERY

ION EXCHANGE

LAND APPLICATION

MANAGEMENT INITIATIVES

MATERIALS SUBSTITUTION

RECYCLING, RECOVERY, AND REUSE

ADSORPTION

CASE STUDY 35

Rexham Corporation does high-technology printing and coating for the photography industry at their Matthews, North Carolina, facility. The have installed a carbon-adsorption-based, air pollution control system which captures vapors from oil-based solvents. The reclaimed solvent is then sold for use in the coatings industry. The recovery system has significantly reduced emissions, and has allowed the company to expand without exceeding air pollution limits.

DISTILLATION

CASE STUDY 9

Rexham Corporation manufactures and prints product labels at their Greensboro, North Carolina, plant. They have installed a distillation unit to receive a solvent blend of normal propyl alcohol and normal propyl acetate. Total cost for the distillation unit and for the structural modifications was approximately $16,000. The unit recovers 85 percent of the solvent in the waste stream, resulting in annual savings of $15,000 in virgin solvent costs, and of $22,800 in hazardous waste disposal costs. The reclaimed alcohol/acetate mixture is reblended with virgin materials to provide the required composition for new batches and for cleanup use. Rexham plans to add a further distillation unit in order to recover the remaining 15 percent of the solvents in sludge wastes.

CASE STUDY 13

American Enka is a nylon yarn production and research facility. They have found it economically profitable and environmentally sound to recycle thier waste isopropyl alcohol solvent in-house, rather than having it recycled by an outside firm. They purchased a used distillation unit, and, with appropriate modifications, American Enka is now saving $90,000 per year. They have also been able to reuse the still bottom residues as an asphalt emulsifier in another product line.

CASE STUDY 18

ICI Americas, Inc., operates an agricultural chemical research and development lab in Goldsboro, North Carolina. As a research facility, it does not produce a large amount of waste. A management initiative, undertaken in 1981, formed an Environmental Compliance Committee to review issues and activities at ICI that affect the environment. The Committee's primary emphasis is on early planning to minimize waste production; and, secondarily, on attempting to recycle or treat the waste, whenever possible, so that none is discharged. Measures recommended by the committee, and then implemented include: housekeeping improvements, and the initiation of waste solvent recovery by distillation. The

Committee's recommendations saved the company an estimated $37,000 in waste disposal costs in 1984.

CASE STUDY 29

Scovill uses 1,1,1-trichloroethane solvent to degrease metal stampings produced in their manufacturing process. Waste solvent is collected in 55-gallon drums and housed in a storage building designed to contain spills, should they occur. The drums are picked up by a company that distills the solvent from the waste and returns it to Scovill for reuse. The recycled solvent costs about half as much as virgn solvent.

CASE STUDY 35

Rexham Corporation does high-technology printing and coating for the photography industry at their Matthews, North Carolina, facility. Off-site distillation of spent solvent mixtures has allowed reuse of methyl-ethyl ketone solvent for washup operations.

ELECTROLYTIC REACTIONS

CASE STUDY 2

Climax Molybdenum mills ore for the production of molybdenum. In order to reduce the amount of water that had to be treated before discharge and to recover the soluble molybdenum from the tailings stream, Climax Molybdenum Company studied the natural drainage system surrounding the tailing impoundment area. They then installed interceptor canals to channel upstream runoff waters from around the tailing system into natural streams. The tailings runoff water is passed through a countercurrent ion exchange unit in order to remove the molybdenum, which is recovered and sold. Other heavy-metal separation from process water is accomplished by the Swift Lectro Clear process, which removes over 90 percent of the iron, manganese, zinc, and copper from the tailings stream by a coagulative electrolytic process. Cyanide, used as a copper depressant in the regrind plant, is 96-percent removed.

CASE STUDY 25

Elkhart Products, Inc., manufactures pipe fittings. Under the guidance of Lancy International, they have modified their bright-dipping and passivating-line wastewater treatment system. It was replaced with a newly developed pickling process that uses sulfuric acid and a proprietary, stabilized hydrogen peroxide. By modifying the equipment operation, Elkhart was able to remove a cyanide dip and chromic acid dip from the process line. They also recover metal with an electrolytic cell, which currently recovers copper at the rate of 1,000 to 1,500 pounds per month.

CASE STUDY 30

GTE manufactures electronic telephone switching equipment. With stricter regulations concerning copper effluents soon to be enforced, GTE, aided by Lancy International, decided to modify the treatment of process water from their copper-etching rinse stream by installing an electrolytic metal-recovery cell. The electrolytic cell is part of a closed-loop system that follows the etching operations on the Electroless copper-plating line. The cell was installed downstream from the rinse tank, where it recovers copper that would have ended up as a landfilled sludge. After the copper ions are removed, the rinse water is recirculate back to the rinse tank. The copper is periodically removed from the cell's cathode and sold as No. 1 copper scrap. In its first year of operation, this system produced $2,000 from the sale of recycled copper scrap, and saved $4,000 in avoided waste disposal costs.

CASE STUDY 33

Digital Equipment Corporation's new, printed wiring board manufacturing operation includes several resource conservation and recovery methods in its process water treatment system. The system, designed by Lancy International, includes metal recovery by electrolytic cell and by evaporation. The electrolytic reaction recovers 92 percent of the dissolved copper from dragout rinses.

CASE STUDY 34

Modine prepares brass parts for radiators by dipping them in a series of chemical solutions. The hydrogen peroxide/sulfuric acid bright dip was causing them several problems, including the buildup of copper sulfate pentahydrate crystals that were then deposited in a hazardous waste landfill. There was, in addition, a significant dragout of copper ions into the subsequent rinse tanks. Rather than incur the initial and long-term landfilling costs for the hazardous wastes, Modine elected to eliminate the formation of the crystals by continuously recirculating the hydrogen peroxide/sulfuric acid bright dip solution through an ion exchange column. The column selectively removes the metal. The purified dip solution is recycled to the bath. Once the ion exchange resin becomes saturated, the copper ions are exchanged again in the regeneration process, and are subjected to an electrolytic reaction to remove them from the solution. The plated copper is sold as No. 1 copper scrap.

CASE STUDY 44

PCA International is a color portrait processing company. An electrolytic metal system is used to recover silver from color negative-film fixing solutions and from the paper bleach-fix solutions. The silver recovery system allows reuse of the film fixer, and the silver chip obtained is about 96-percent pure.

CASE STUDY 45

American Fotokemi, Inc., established a central processing plant for recovering silver from films and spent fixer solutions used in film processing. Conventional approaches to silver recovery in the film processing industry have located canister and electrolytic recovery equipment at each individual film-processing site. This decentralized approach results in a less than optimal recovery of silver. American Fotokemi has developed chemical treatment processes, used in conjunction with electrolytic recovery, that increase silver recovery to nearly 100 percent -- an increase of 10 to 15 percent from that achieved by conventional, decentralized technology. Furthermore, other toxic metals, which would otherwise be discharged as pollutants by the small, independent silver-recovery operations, are also recovered. The high efficiency of this centralized recovery method has increased the dollar/silver return for everyone who has converted to it from in-house canister and/or conventional electroplating recovery.

EQUIPMENT DESIGN AND OPERATION

CASE STUDY 3

California Canners and Growers (CCG) cans produce. They previously operated their atmospheric cookers one-half to two-thirds full of water. Any steam that did not condense in the water escaped through the can in-feed and discharge ports and was wasted. To utilize this wasted food-processing energy, CCG, aided by Pardee Engineering, devised a modification to the existing equipment design which allowed the retention of steam in the cooker. They also removed the water in the cooker so that the steam now condenses directly on the cans. This has resulted in considerable energy, water, and labor savings.

CASE STUDY 6

Hampshire Hosiery manufactures women's pantyhose. They replaced an existing oil-fired boiler system for steam generation with a wood-fired system capable of using sawdust from nearby lumbermills as fuel. The new boiler operation saves the firm $200,000 per year.

CASE STUDY 10

Exxon Chemical installed floating roofs over its tanks of volatile solvent. This has prevented the loss of more than $1.5 million worth of solvents. Concurrently, they installed conservation vents on the plant's large hydrocarbon tanks, which reduced air emissions from the tanks by 30 to 75 percent. In addition to reducing air pollution, these two equipment design modifications have conserved a significant amount of resources. A management initiative created a stewardship program to reduce fluctuations in the wastewater treatment system. Sampling stations were established to monitor concentrations of organic chemicals; as a result, operators can trace any fluctuations to their origin and quickly remedy the problem. The information gained from the monitoring is also used to charge each process with a portion of the costs of the wastewater treatment system, and to set targets for waste reduction for each process. This program resulted in a 75-percent reduction in the quantity of organic wastes entering the wastewater treatment plant.

CASE STUDY 12

Borden Chemical Company reduced the amount of organic phenol resins they discharged by 95 percent, mainly through three process modifications:

1) Phenolic resins are passed through a filtration process that removes the large resinous particles. Previously, the filtered materials were disposed of by rinsing them into the wastewater treatment system. They are now rinsed into large tanks and recycled into the process as raw material.

2) Reactor vessels were cleaned by filling them with water. This produced a large amount of phenolic resin waste that had to be processed through the wastewater treatment system. New equipment operation procedure has reduced the initial rinse to only 5 percent of the previous volume, resulting in a more concentrated solution of phenolic resins. This solution is stored and recycled into the process line as a raw material. The reactor vessels are then rinsed a second time. The waste stream from this rinse contains a low concentration of phenolic resins.

3) The Borden Company has made employees aware that small amounts of chemical waste can strongly affect overall waste generation. As a result, management and workers have identified ways of reducing wastes.

CASE STUDY 14

Riker Laboratories was coating medicine tablets, using several organic solvents. Consequently, they had the potential for exceeding air pollution limits set by a new, California environmental regulation. A water-based solvent was developed to replace the organic solvent, and different spray equipment was installed in order to handle the new solvent. This combination of resource substitution of the water-based solvent for the organic solvent, and of equipment design modification saves $15,000 per year in solvent costs, and prevents the production of 24 tons of air pollution per year. Subsequently, the company determined that $180,000 in pollution control equipment for organic solvent vapors was no longer required.

CASE STUDY 16

USS Chemicals has reduced the organic chemical emissions from their phenol manufacturing processes by a series of equipment design modifications including:

EQUIPMENT DESIGN AND OPERATION, CONTINUED

1) Adding a refrigerated condenser unit, in combination with a resin adsorption system, at a cost of $5,000. This reduced emissions by 80 percent, and saves 237,000 pound of cumene per year. The cumene savings is worth $60,000 per year.

2) Adding a scrap condenser that returns the emissions to the process. This saved 400,000 pound of cumene, worth $100,000, in its first year of operation.

3) Adding floating roofs to acetone storage tanks, and an adsorber unit for acetone recovery.

Modification #2 arose from employee ideas submitted to the USS Chemicals' "Suggestions for Cost Reduction" (SCORE). SCORE is a management initiative which provides plant employees with a direct financial incentive for identifying cost-cutting practices; the program rewards them with a percentage of the savings that have resulted from the implementation of their ideas. Employees have received a total of $70,000 for successful suggestions that have saved the firm $500,000 in costs.

CASE STUDY 19

Daly-Herring Company manufactures pesticides and insecticides. They altered their dust collection equipment so that waste streams coming from various production areas are now collected separately, rather than mixed in a single baghouse. Th collected materials are no longer contaminated by alternate waste streams, and each material is recycled back to the process where it originated. The firm has eliminated over $9,000 in annual disposal costs, and estimates that the recovered material is wort more than $2,000 per year.

CASE STUDY 24

Stanadyne, Inc., manufactures plumbing products that require cleaning and electroplating operations which generate hazardous wastes. As part of a comprehensive review of their processes undertaken to reduce or eliminate waste production, several equipment design and operation modifications were made. The company found they could operate their copper- and chrome-electroplating bath equipment at a lower cyanide concentration, saving both raw material and sludge handling costs. Other changes reduced dragout volumes and rack recoating costs, while also improving plating efficiency. Tanks were installed, before the rinsing operation, to recycle valuable metals back to the plating baths. A change in the operation of plating bath filtration equipment drastically reduced solution losses, and resulted in a 93-percent cost reduction in filter media. A filter press operation for sludge dewatering has minimized the cost of sludge transport and disposal. A new wastewater filtration system has improved the quality of treated process water.

CASE STUDY 25

Elkhart Products, Inc., manufactures pipe fittings. Under the guidance of Lancy International, they have modified their bright-dipping and passivating-line wastewater treatment system. It was replaced with a newly developed pickling process that uses sulfuric acid and a proprietary, stabilized hydrogen peroxide. By modifying the equipment operation, Elkhart was able to remove a cyanide dip and chromic acid dip from the process line. They also recover metal with an electrolytic cell which recovers copper at the rate of 1,000 to 1,500 pounds per month.

CASE STUDY 26

Pioneer was using a single-pass system to treat their plating wastewaters. Because the system had no large end-tank to correct minor upsets in the system, concentrations of metals and other inorganic chemicals in the effluent regularly exceeded water quality regulations. The effluents were killing many species of plants and animals in the receiving marsh. Pioneer solved this problem by removing the old, single-pass process water treatment system and installing a closed-loop batch system manufactured by Zerpol. As a result of the new equipment design, the company now uses only 500 gallons of water per day (makeup due to evaporation), rather then the previous 12,000 gallons per day, and has experienced a 20-percent reduction in the amount of sludge produced. Plans are being made to sell the sludge to a company for use as a filler in the manufacture of concrete products.

CASE STUDY 28

Emerson Electric Company manufactures stationary power tools, which require both metal finishing and painting operations. They installed an automated metal-electroplating system to replace their manual operation, and increased annual productivity by $200,000. The substitution of a water-based paint system, which uses anodic electrostatic immersion, for the existing organic solvent-based paint system improved annual productivity by $1 million. The water-based system allows 99.5-percent recovery and reuse of paint. Hazardous waste disposal costs have been reduced from $10,000 per year to only #300 per year. Equipment modifications in a die-cast sump and pumping facility allowed the recovery of 65 gallons of waste oil per week. A steam cleaner was purchased for $5,000, and degreasing is accomplished by using an alkaline detergent and steam, rather than chlorinated organic solvents. Adding chlorinators to the plant's cooling towers eliminated the influx of zinc chromate algicide into the storm drains. Finally, a computerized energy management system was installed in the plant, which saves the company approximately $147,000 per year.

CASE STUDY 31

Data General Corporation generates several hazardous waste streams during the manufacture of printed circuit boards. Their hazardous waste management modifications have focused on selling their wastes for reuse or recovery. Several equipment modifications have made some of their wastes more marketable. Modification of their existing water purification filters allowed quicker, easier cleaning. When coupled with a process chemistry change, the resulting sludge could be successfully processed for copper recovery. An air blow-down line was added to their existing filter press to increase the sludge solids content, and reusable polishing filters were developed for treating the water effluent.

CASE STUDY 32

3M's microelectronics plant makes flexible electronic circuits from copper sheeting. Before use, the metal was cleaned by spraying it with ammonium persulfate, phosophoric acid, and sulfuric acid. This process produced hazardous wastes that were expensive to manage properly. An alternative equipment design was introduced to replace the chemical cleaning process. The new process features a machine with rotating brushes that mechanically clean the metal with pumice. The fine, abrasive pumice leaves a sludge that is not hazardous and that can be placed in a conventional sanitary landfill. This new method reduced 3M's production of hazardous waste by 40,000 pounds a year.   It also saved $15,000 in its first year by reducing raw material, disposal, and labor costs.

CASE STUDY 37

Duke Power Company is one of the major electric utilities in North Carolina.   The Company assembled a team of staff members to pursue low-level radioactive waste reduction strategies.   Improved compactor equipment was installed to reduce the volume of wastes requiring disposal.   Decontamination equipment was purchased to allow cleanup and reuse rather than disposal of contaminated tools and equipment.   Existing evaporator equipment, originally designed for processing non-recyclable waste, was extensively modified to process recyclable liquid wastes.   The conversion improved evaporator performance and minimized boron and tritium release to the environment.   Filtration equipment was modified to improve filter performance and to minimize replacement and disposal costs. The filter modifications required a $60,000 investment, but they have saved $200,000 per year.

CASE STUDY 41

The chemical and coordinate bond and adsorbtion system (CCBA) is a process for treating a municipal waste (raw sewage).   The equipment design and operation is based on chemical processes rather then traditional biological decomposition and stabilization.   Common clay, alum, and polyacrylic acid are added to the sewage influent.   The chemical interaction of these compounds with the dissolved and solid sewage waste is effective in removing of 91 percent of the suspended solids, and 88 percent of the biological oxygen demand from the waste water.   Remaining organics in the plant effluent are water-soluble and easily biodegradable.   Sludge, resulting from the cleaning of the influent wastewater, is mixed with sufficient clay and then baked in a kiln to produce a lightweight aggregate suitable for reuse in the building industry.   Additional processing of the treated wastewater permits significant removal of nutrient wastes.   The processing facility is significantly less expensive to construct than conventional biological sewage treatment facilities, and is not sensitive to toxic constituents in the wastewater.   An additional benefit accruing from this process is the ability to safely treat hazardous inorganic and organic chemicals in the sewage sludge by incinerating them at elevated temperatures; the residues are subsequently entrained in a ceramic matrix.

CASE STUDY 46

Vulcan Automotive Equipment Ltd. remanufactures used automotive engines. The cleaning process was modified to replace the inorganic caustic cleanser with a high-velocity "aluminum shot" system. This modification in the cleanser, and in associated equipment design, resulted in substantial savings to the company due to decreased costs of raw materials and labor, and to avoided waste management costs. The new, aluminum shot system increases productivity and improves the final product. All by-products of the new process are recycled.

FILTRATION

CASE STUDY 12

Borden Chemical Company reduced the amount of <u>organic</u> phenol resins it discharged by 95 percent, mainly through three process modifications, one of which one was filtration.

Phenolic resins are passed through a filtration process that removes the large resinous particles. Previously, the filtered materials were disposed of by rinsing them into the wastewater treatment system. They are now rinsed into large tanks and recycled into the process as raw material.

CASE STUDY 28

Emerson Electric Company manufactures stationary power tools which require both metal finishing and painting operations. The addition of a $65,000 oil skimmer and ultrafiltration system to their existing wastewater treatment system has allowed recovery of oil and cleaning solution. The unit has cut water treatment costs by $10,000 per year.

CASE STUDY 31

Data General Corporation generates several hazardous waste streams during the manufacture of printed circuit boards. Process water from circuit board fabrication, assembly, and testing is treated by water purification system which uses cross-flow filtration. The filtration effluent is disposed of by application to forestland owned by the company.

HOUSEKEEPING

CASE STUDY 18

ICI Americas, Inc., operates an agricultural chemical research and development laboratory in Goldsboro, North Carolina. The company's primary emphasis is on early planning to minimize waste production, and, secondarily, on attempting to recycle or treat the waste, whenever possible, so that none is discharged. Housekeeping measures and improvements are important because ICI does not generate large quantities of wastes, and cannot justify large capital expenditures for their management. Housekeeping innovations and solvent recovery by distillation saved the company an estimated $37,000 in waste disposal costs in 1984.

CASE STUDY 24

Stanadyne, Inc., is a manufacturer of plumbing products which require electroplating and cleaning operations that generate hazardous waste. When the company undertook a comprehensive review of their processes in order to eliminate unnecessary waste, one outcome of their analysis was improved housekeeping. Preinspection of parts, to eliminate obvious rejects before plating, and a leak elimination program were instituted to reduce plating solution and dragout wastes.

CASE STUDY 28

Emerson Electric Company manufactures stationary power tools requiring both metal finishing and painting operations. Several new housekeeping procedures were instituted to reduce waste generation. They included the cessation of unauthorized dumping of waste chemicals in the on-site wastewater treatment system, and improved drum storage and handling procedures for wastes stored on-site. These changes eliminated the risk of environmental damage from high concentrations of metals in the treated effluent and from improperly protected wastes stored on-site.

CASE STUDY 35

Rexham Corporation does high-technology printing and coating for the photographic industry at their facility in Matthews, North Carolina. A new housekeeping procedure, in which cleaning solvent is segregated by the color and type of ink contaminant, has allowed almost 100-percent reuse of the solvent.

CASE STUDY 37

Duke Power Company is one of the major electric utilities in North Carolina. A team of staff members was assembled to develop comprehensive strategies for low-level radioactive waste reduction. The resulting program included an effort to improve housekeeping activities. Compactible materials suspected of radioactive contamination were segregated in different areas in the plant as "potentially contaminated" and "not contaminated". Since much of the material was free of radioactivity, this procedure significantly reduced the amount of hazardous waste requiring disposal.

INCINERATION/HEAT-ENERGY RECOVERY

CASE STUDY 6

Hampshire Hosiery manufactures women's pantyhose. They have installed a heat exchanger to capture heat from spent dye solutions and to preheat subsequent dye operations. In addition to energy savings, the exchanger system lowers the temperature of the effluent entering the municipal sewer system from 130 degrees F to 70 degrees F; this improves treatment performance.

CASE STUDY 7

United Globe generates two major waste streams during their furniture manufacturing operations. They purchased an incinerator and used their treated waste streams as fuel for producing steam and heat. Solids from the ragwash and spraybooth process water are thickened using dissolved air flotation, mixed with woodchips, dried, and fed to the incinerator. Semi-solid and liquid finishing wastes, including high-BTU solvents, are also fed to the incinerator. The incinerator has reduced United Globe's waste volume by 95 percent, and the ash produced is not hazardous. Discharge loading to the municipal sewer system has decreased dramatically, as have hazardous waste transport and disposal costs.

CASE STUDY 15

The Coty Division of Pfizer, Inc., has developed a method for recovering heat from waste hydro-alcoholic products. The ignitable wastes are mixed with fuel oil and burned to produce steam and hot water. The company saves more than 1,000 gallons of fuel oil a year, and has eliminated the costs of hazardous waste disposal.

CASE STUDY 35

The Rexham Corporation facility in Matthews, North Carolina, laminates and coats specialized paper, film, and foil substrates for the photography industry. A management initiative to develop a comprehensive waste management plan for the firm has resulted in several pollution prevention modifications. Off-site distillation of spent <u>solvent</u> mixtures has allowed the reuse of methyl-ethyl-ketone solvent for washup operations. Methyl-ethyl-ketone vapors are removed from the air and are incinerated for heat recovery.

CASE STUDY 37

Duke Power Company is one of the major generators of electric power in North Carolina. Their efforts to reduce the amount of polychlorinated biphenyl (PCB) wastes requiring hazardous waste disposal are part of an overall program to phase out PCB contamination from their electrical distribution system. Insulating fluid contaminated with PCBs is burned as a supplemental fuel for heat recovery at a steam-electric generating station. The PCBs are totally destroyed, and the Company has saved about $84,000 in hazardous waste disposal costs in each of the past few years.

ION EXCHANGE

CASE STUDY 2

Climax Molybdenum Company mills ore to produce molybdenum. In order to reduce the amount of water that had to be treated before discharge and to recover the soluble molybdenum from the tailings stream, Climax Molybdenum Company studied the natural drainage system surrounding the tailing impoundment area. They then installed interceptor canals to channel upstream runoff waters from around the tailing system into natural streams. The tailings runoff water is passed through a countercurrent ion exchange unit in order to remove the molybdenum, which is recovered and sold. Other heavy metal separation from process water is accomplished by the Swift Lectro Clear process, which removes over 90 percent of the iron, manganese, zinc, and copper from the tailings stream by a coagulative electrolytic process. Cyanide, used as a copper depressant in the regrind plant, is 96-percent removed.

CASE STUDY 34

Modine prepares brass parts for radiators by dipping them in a series of chemical solutions. The hydrogen peroxide/sulfuric acid bright dip was causing them several problems, including the buildup of copper sulfate pentahydrate crystals that were then deposited in a hazardous waste landfill. There was, in addition, a significant dragout of copper ions into the subsequent rinse tanks. Rather than incur the initial and long-term landfilling costs for the hazardous metal wastes, Modine elected to eliminate the formation of the crystals by continuously recirculating the hydrogen peroxide/sulfuric acid bright dip solution through an ion exchange column. The column selectively removes the metal. The purified dip solution is recycled to the bath. Once the ion exchange resin becomes saturated, the copper ions are exchanged again in the regeneration process, and are subjected to an electrolytic reaction to remove them from the solution. The plated copper is sold as No. 1 copper scrap.

CASE STUDY 44

PCA International is a color portrait processing company. They installed an ion exchange system which removes bromide from color developer solution and allows regeneration and reuse of the developer. The system has allowed 99-percent reuse of the color developer solution.

LAND APPLICATION

CASE STUDY 4

Dunbar Foods processes canned peppers and sweet potatoes. They developed the first land application system for industrial wastewater in North Carolina. Process water containing vegetable wastes is applied by spray irrigation to grassland owned by the company.

CASE STUDY 5

Miller Brewing Company is a beer manufacturer in Eden, North Carolina. Food processing wastes generated during the brewing operation are treated and made available for land application. The wastes have value for liming and as a nitrogen supplement. The reuse of this previously landfilled material has eliminated disposal costs for Miller, provided a virtually cost-free substitute for standard agricultural liming material and fertilizer, and preserved valuable landfill capacity.

CASE STUDY 31

Data General Corporation manufactures printed circuit boards for computers. Process water containing high concentrations of heavy metals is passed through a filter-based water purification system, and is then disposed of by land application to forestland owned by the company. This was the first site in North Carolina for land application of pretreated metalplating waste.

MANAGEMENT INITIATIVES

CASE STUDY 1

The Choplin Family Farm raises swine and various field crops. In 1982, the Choplin farm saved $8,000 in fertilizer costs; they reduced organic material losses by 1,670 percent, total nitrogen losses by 490 percent, and total phosphorus losses by 525 percent. The management initiative to incorporate conservation practices in the farm operation resulted in corn crop yields that were 300 percent above the county average.

CASE STUDY 8

West Point Pepperell is one of the major textile manufacturers in the nation. They established a Toxic Chemicals Committee, in 1975, to review products in use or under consideration, and to evaluate them according to a wide range of health and environmental impact criteria. The company believes that the Committee has helped them to minimize their production of hazardous waste and its accompanying costs and environmental impacts.

CASE STUDY 16

USS Chemicals has reduced the organic chemical emissions from their phenol manufacturing processes by a series of equipment design modifications, many of which arose through employee ideas submitted to USS Chemicals' "Suggestions for Cost Reduction" (SCORE). SCORE is a management initiative which provides plant employees with a direct financial incentive for identifying cost cutting practices; the program rewards them with a percentage of the savings that have resulted from the implementation of their ideas. Employees have received a total of $70,000 for successful suggestions that have saved the firm $500,000 in costs.

CASE STUDY 18

ICI Americas, Inc., operates an agricultural chemical research and development laboratory in Goldsboro, North Carolina. The company's primary emphasis is on early planning to minimize waste production, and, secondarily, on attempting to recycle or treat the waste, whenever possible so that there none is discharged. Housekeeping measures and improvements are important because ICI does not generate large quantities of wastes, and cannot justify large capital expenditures for their management. Housekeeping innovations and solvent recovery by distillation saved the company an estimated $37,000 in waste disposal costs in 1984.

CASE STUDY 24

Stanadyne, Inc., manufactures plumbing parts. Their operation requires cleaning and electroplating processes that generate hazardous wastes. The management at Stanadyne has a comprehensive program for reducing waste production; they believe the key to the success of the program is employee education and participation. Representatives from Stanadyne have participated in numerous state and local programs to educate other industries and the general public about hazardous wastes and pollution prevention.

CASE STUDY 27

Deere and Company reduced its hazardous waste by 80 percent in volume and over 99 percent in level of risk through two key actions:

1) Implementation of a comprehensive hazardous waste management initiative at each manufacturing unit.

2) Construction of an on-site hazardous liquid waste treatment facility that reclaims waste organic oil compounds and metals from process water, and produces a non-hazardous sludge. The recovered oil is sold to an oil recycling firm or is reused for machining processes.

CASE STUDY 28

Emerson Electric Company manufactures stationary power tools which require both metal finishing and painting. The president and vice president of operations developed and led a broad- based waste management program. Daily log sheets are maintained to insure that the stated procedures are followed. The Company also maintains an incentive awards program to encourage employee suggestions for cost savings and waste reduction.

CASE STUDY 29

Scovill Corporation manufactures small appliances. Their cleaning and degreasing operations generate hazardous wastes. They have instituted an incentive awards program to encourage employee ideas for cost savings and waste reduction. Monetary awards are given each year to the team of employees making the greatest contributions in these areas.

CASE STUDY 31

Data General Corporation manufactures printed circuit boards for computers. In 1981, the company initiated a management policy stating that landfill disposal of wastes should be used only as a last resort. In order to reduce their reliance on landfilling, primary emphasis was placed on the sale of spent chemicals and sludges, rather than on reducing or eliminating the production of these wastes. Data General's program to reduce wastes has included marketing activities to find buyers for their wastes, operational revisions, and development of new process chemistry to make their wastes more salable.

CASE STUDY 35

The Rexham Corporation facility in Matthews, North Carolina, laminates and coats specialized paper, film, and foil substrates for the photography industry. A management initiative to develop a comprehensive waste management plan for the firm has resulted in several pollution prevention modifications. Off-site distillation of spent solvent mixtures has allowed the reuse of methyl-ethyl-ketone solvent for washup operations. Methyl-ethyl-ketone vapors are removed from the air and are incinerated for heat recovery.

CASE STUDY 37

Duke Power Company is one of the major electric utilities in North Carolina. A team of staff members has been assembled at each of Duke's power stations to oversee waste management activities. At the McGuire Nuclear Station, a special section in the Company's employee handbook stresses the importance of reducing wastes through a variety of activities. Management policy emphasizes that wastes should be reduced as much as is operationally possible, not just to the point of satisfying predetermined legal limits.

MATERIAL SUBSTITUTION

CASE STUDY 9

Rexham Corporation manufactures and prints product labels at their Greensboro, North Carolina, plant. They found that replacing alcohol/acetate-borne inks with waterborne inks significantly reduced spent <u>solvent</u> volumes and air emissions.

CASE STUDY 14

Riker Laboratories was coating medicine tablets, using several organic solvents. Consequently, they could potentially exceed air pollution limits set by a new, California environmental regulation. A water-based <u>solvent</u> was developed to replace the organic solvent, and different spray equipment was installed in order to handle the new solvent. This combination of the resource substitution of water-based solvent for organic solvent, and of equipment design modification saves $15,000 per year in solvent costs, and prevents the production of 24 tons of air pollution per year. Subsequently, the company determined that $180,000 in pollution control equipment for organic solvent vapors was no longer required.

CASE STUDY 17

IAC manufactures trace-element compounds for fertilizer formulations. They use materials substitution to reduce their raw material costs. Suitable <u>inorganic chemical</u> wastes are received from firms that would otherwise pay to landfill them.

CASE STUDY 24

Stanadyne, Inc., manufactures plumbing products. Their cleaning and electroplating operations generate hazardous wastes. The substitution of a sulfuric/peroxide bright-dipping solution for a chromic acid-based bright dip reduced the amount of hazardous sludge produced by this process.

MATERIAL SUBSTITUTION, CONTINUED

CASE STUDY 29

Scovill, Inc., manufactures small appliances and uses 1,1,1-trichloroethane cleaning solvent. They found that, by substituting water-soluble synthetic cleaner for some of their degreasing applications, they could reduce the volume of spent organic solvent they generated. The substitution has saved the company approximately $12,000 per year.

RECYCLING, RECOVERY, AND REUSE

CASE STUDY 1

The Choplin Family Farm raises swine and various field crops. In 1982, the Choplin farm saved $8,000 in fertilizer costs by spraying swine wastes on crops with a standard irrigation gun. The reuse of these agricultural wastes and the incorporation of conservation practices in the farm operation resulted in corn crop yields that were 300 percent above the county average.

CASE STUDY 4

Dunbar Foods, Inc., processes canned peppers and sweet potatoes. The development of a spray irrigation system for the land application of their waste process water has allowed them to reuse the water for irrigation and fertilization. The system also eliminated disposal to the municipal sewer system.

CASE STUDY 5

Miller Brewing Company generates activated waste sludge as a by-product of their beer brewing operation. The treated sludge is made available for land application as a liming agent and nitrogen supplement. The reuse of this previously landfilled material has eliminated disposal costs for Miller and has preserved valuable landfill capacity.

CASE STUDY 11

Allied Corporation manufactures chemicals. A $4.5 million investment in recycling equipment is paying off at the rate of $1 million a year for Allied Corporation. Allied's Metropolis, Illinois, plant recovers 8,000 tons of calcium fluoride annually. The inorganic chemical is then used as a raw material in anhydrous hydrofluoric acid production at another facility. Since 1982, over 1,000 cubic yards of hazardous calcium fluoride wastes have been recycled monthly, saving about $300,000 a year in disposal and storage costs alone. The process also recovers about 1,000 tons of lime annually.

RECYCLING, RECOVERY, AND REUSE, CONTINUED

CASE STUDY 12

Borden Chemical Company reduced the amount of organic phenol resins it discharged by 95 percent, mainly through three process modifications:

1) Phenolic resins are passed through a filtration process that removes the large resinous particles. Previously, the filtered materials were disposed of by rinsing them into the wastewater treatment system. They are now rinsed into large tanks and recycled into the process as raw material.

2) Reactor vessels were cleaned by filling them with water. This produced a large amount of phenolic resin waste that had to be processed through the wastewater treatment system. New procedures for equipment operation have reduced the initial rinse to only 5 percent of the previous volume, resulting in a more concentrated solution of phenolic resins. This solution is stored and then recycled into the process line as a raw material. The reactor vessels are then rinsed a second time. The waste stream from this rinse contains a low concentration of phenolic resins.

3) The Borden Company has made employees aware that small amounts of chemical waste can strongly affect overall waste generation. As a result, management and workers have identified ways of reducing wastes.

CASE STUDY 13

American Enka is a nylon yarn production and research facility. They have found it economically profitable and environmentally sound to recycle their waste isopropyl alcohol solvent in-house rather than having it recycled by an outside firm. They purchased a used distillation unit and, with appropriate modifications, American Enka is now saving $90,000 per year. They also reuse the still bottom residues as an asphalt emulsifier in another product line.

CASE STUDY 17

IAC manufactures trace-element compounds for fertilizer. By substituting suitable inorganic waste chemicals received from other firms for some of their virgin raw materials, they have they have been able to reduce their production costs and prevent the landfilling of useable material.

296

RECYCLING, RECOVERY, AND REUSE, CONTINUED

CASE STUDY 18

ICI Americas, Inc., operates an agricultural chemical research and development lab in Goldsboro, North Carolina. A management initiative was undertaken, in 1981, to form an Environmental Compliance Committee to review issues and activities at ICI that affect the environment. Primary emphasis is on early planning to minimize waste production, and, secondarily, on attempting to recycle or treat the waste, whenever, possible so that none is discharged. The Committee's recommendations saved the company an estimated $37,000 in waste disposal costs in 1984.

CASE STUDY 19

Daly-Herring Company manufactures organic pesticides and insecticides. By segregating dust waste streams, they have been able to recycle each dust-type back to the process where it originated. The firm has eliminated over $9,000 in annual disposal costs, and estimates that the recovered material is worth more than $2,000 per year.

CASE STUDY 20

Texasgulf manufactures phosphate products. In designing their new, defluorinated phosphate plant to meet EPA regulations pertaining to the release of fluorides into the air and water, they decided to implement a closed-loop process water system to remove inorganic fluorides from the discharge stream. Other alternatives considered included releasing the partially treated wastewater to the Saltville, Virginia, municipal sewage system, or discharging it into the nearby Holton river. Benefits of their decision to implement a closed-loop rather than a single-pass system to recycle process water include:

1) Savings of approximately 280,000 gallons of water per day.

2) Combined annual savings from water purchases and treatment of $930,000 per year.

3) Compliance with all effluent guidelines without a NPDES permit, since there is no industrial discharge.

CASE STUDY 22

Florida Steel Corporation produces steel in two electric-arc furnaces. Metal dust collected from the furnace fumes is sold to a zinc smelter for metal recovery. This practice has reduced their disposal costs by almost 50 percent, and prevents burial of a valuable, recoverable resource.

CASE STUDY 24

Stanadyne, Inc., is a manufacturer of plumbing products. Their electroplating and cleaning operations generate hazardous waste. The company altered their nickel electroplating process so that dragout is captured in still tanks before the rinsing step is begun. The dragout is returned to the plating tanks, thus saving valuable chemicals and reducing the treatment and disposal costs of removing high concentrations of metals from the rinse water. The recovered nickel saves Stanadyne $4,200 per year and reduces the nickel sludge volume by 9,500 pounds per year.

CASE STUDY 25

Elkhart Products, Inc., manufactures pipe fittings. Under the guidance of Lancy International, they have modified their bright-dipping and passivating-line wastewater treatment system. It was replaced with a newly developed pickling process that uses sulfuric acid and a proprietary, stabilized hydrogen peroxide. By modifying the equipment operation, Elkhart was able to remove a cyanide dip and chromic acid dip from the process line. They also recover metal with an electrolytic cell which currently recovers copper at the rate of 1,000 to 1,500 pounds per month.

CASE STUDY 26

Pioneer was using a single-pass system to treat their plating wastewaters. Because the system had no large end-tank to correct minor upsets in the system, concentrations of metals and other inorganic chemicals in the effluent regularly exceeded water quality regulations. The effluents were killing many species of plants and animals in the receiving marsh. Pioneer solved this problem by removing the old, single-pass process water treatment system and installing a closed-loop batch system manufactured by Zerpol. As a result of the new equipment design, the company now uses only 500 gallons of water per day (makeup due to evaporation), rather then the previous 12,000 gallons per day, and has reduced the amount of sludge produced by 20 percent. Plans are being made to sell the sludge to a company for use as a filler in the manufacture of concrete products.

CASE STUDY 27

Deere and Company reduced its hazardous waste by 80 percent in volume and over 99 percent in level of risk through two key actions:

1) Implementation of a comprehensive hazardous waste management initiative at each manufacturing unit.

2) Construction of an on-site hazardous liquid waste treatment facility that reclaims waste organic oil compounds and metals from process water, and produces a non-hazardous sludge. The recovered oil is sold to an oil recycling firm or is reused for machining processes.

CASE STUDY 30

GTE Sylvania manufactures microelectronic telephone switching equipment. With stricter regulations concerning copper effluents soon to be enforced, GTE, aided by Lancy International, decided to modify the process water treatment from their copper-etching rinse stream by installing an electrolytic metal recovery cell. The electrolytic cell is part of a closed-loop system that follows the etching operations on the Electroless copperplating line. The cell was installed downstream from the rinse tank, where it recovers copper that would have ended up as a landfilled sludge. After the copper ions are removed, the rinse water is recirculated back to the rinse tank. The copper is periodically removed from the cell's cathode and sold as No. 1 copper scrap. In its first year of operation, this system produced $2,000 from the sale of the recycled copper scrap, and saved $4,000 in avoided waste disposal costs.

CASE STUDY 33

Digital Equipment Corporation's new printed wiring board manufacturing operation includes several resource conservation and recovery methods in their process water treatment system. The system, designed by Lancy International, includes metal recovery by electrolytic cell and by evaporation. The electrolytic reaction recovers 92 percent of the dissolved copper from dragout rinses, and the atmospheric evaporator recovers 95 percent of the chromic acid dragout. The chromic acid is recycled into the etch process line. The company also uses a lime slurry by-product from a local acetylene gas manufacturer as an acid neutralizer.

CASE STUDY 34

Modine prepares brass parts for radiators by dipping them in a series of chemical solutions. The hydrogen peroxide/sulfuric acid bright dip was causing them several problems, including the buildup of copper sulfate pentahydrate crystals that were then deposited in a hazardous waste landfill. There was, in addition, a significant dragout of copper ions into the subsequent rinse tanks. Rather than incur the initial and long-term landfilling costs for the hazardous wastes, Modine elected to eliminate the formation of the crystals by continuously recirculating the hydrogen peroxide/sulfuric acid bright dip solution through an ion exchange column. The column selectively removes the metal. The purified dip solution is recycled to the bath. Once the ion exchange resin becomes saturated, the copper ions are exchanged again in the regeneration process, and are subjected to an electrolytic reaction to remove them from the solution. The plated copper is sold as No. 1 copper scrap.

CASE STUDY 35

Rexham Corporation does high-technology printing and coating for the photography industry at their Matthews, North Carolina, facility. The segregation of waste toluene, used for print and roller cleanup, by color and type of ink contaminant has allowed almost complete reuse of the solvent.

CASE STUDY 36

Carolina Power and Light Company is a major generator of electric power in North Carolina. They have been selling waste fly ash and bottom ash from their coal-fired plants since 1973. The Company uses another firm, Monier Resources, Inc., to oversee the marketing of the ash. Prior to this arrangement, all of the ash was sluiced to on-site ash ponds for permanent disposal. The Company has now reduced its disposal costs and prevented the burial of useable material.

CASE STUDY 37

Duke Power Company is one of the major electric utilities in North Carolina. Since the late 1960s, they have been marketing waste fly ash and bottom ash from their coal-fired power plants. The ash has been used for concrete production, asphalt filler, and plastics filler. The Company is exploring other uses for the ash, particularly the possibility of using it as structural backfill or to stabilize roadbases.

CASE STUDY 38

Natural Power is a small generator of electricity operating from the Gresham Lake Landfill in Raleigh, North Carolina. They have implemented a landfill gas recovery system for converting biogas to electric power, which is then sold to the local utility. The gas is collected, filtered, and fed directly into two diesel generators modified to use a methane fuel source. The electrical energy generated at the landfill is metered and fed into the local utility's transmission lines.

CASE STUDY 39

A system for composting municipal waste sludge is planned for the Hickory/Newton/Conover Complex in Catawba County, North Carolina. The facility is in the final design phase, and is scheduled for construction in 1986. Reusing waste sludge as composting material is expected to save the county as much as $827,000 in annual disposal costs.

CASE STUDY 40

The municipal wastewater treatment facility in Morganton, North Carolina, implemented the first, full-scale, sludge composting operation in the state. They purchase bark as a bulking material, and compost the sludge using the static-pile method. The compost is sold to the public, and the response has been so favorable that the city has been unable to keep up with the demand. They have eliminated the need to landfill the waste sludge, and have created a system to reuse an otherwise undesirable material.

CASE STUDY 41

The chemical and coordinate bond and adsorbtion system (CCBA) is a process for treating a municipal waste (raw sewage). The equipment design and operation is based on chemical processes rather then traditional biological decomposition and stabilization. Common clay, alum, and polyacrylic acid are added to the sewage influent. The chemical interaction of these compounds with the dissolved and solid sewage waste is effective in removing of 91 percent of the suspended solids, and 88 percent of the biological oxygen demand from the waste water. Remaining organics in the plant effluent are water-soluble and easily biodegradable. Sludge, resulting from the cleaning of the influent wastewater, is mixed with sufficient clay and then baked in a kiln to produce a lightweight aggregate suitable for reuse in the building industry. Additional processing of the treated wastewater permits significant removal of nutrient wastes. The processing facility is significantly less expensive to construct than conventional biological sewage treatment facilities, and is not sensitive to toxic constituents in the wastewater. An additional benefit accruing from this process is the ability to safely treat hazardous inorganic and organic chemicals in the sewage sludge by incinerating them at elevated temperatures; the residues are subsequently entrained in a ceramic matrix.

CASE STUDY 42

Chevron USA, Inc., maintains a petroleum product warehouse in Louisville, Kentucky. A shipping-drum reconditioning facility, where used drums are cleaned and repainted for reuse, is operated on-site. Prior cleaning practices generated a complex hazardous waste stream, composed of caustic, oil, water, and paint which cost Chevron $50,000 per year to manage. A high-pressure, hot-water cleaning system was installed to replace the caustic cleaning system. The new system is a closed-loop wastewater process which reuses the filtered cleaning water and ensures the recycling of oil cleaned from the drums. A substitution of oil-based paints with a water-base process, using state-of-the-art overspray filtration equipment, conserves paint and eliminates the generation of hazardous waste paint residue. The total savings to Chevron from these two modifications exceeds $80,000 per year, after the 3-year payback period.

CASE STUDY 43

A film developing unit at 3M's Electronics Products Division was discharging wastewater contaminated with 1,1,1-trichloroethane. In order to recover used solvent and to continue in compliance with process water discharge regulations,

3M installed a decanter system that provides gravity separation
of the solvent from the water. This modification, saved the
company $12,000 in its first year of operation by reducing the
amount of new solvent and makeup water required for the
developing unit.

CASE STUDY 44

PCA International is a color portrait processing company.
They have installed several new equipment systems which allow
them to reuse large volumes of process water and process
chemicals. A reverse osmosis, water purification system permits
the recycling of 90 percent of the process water, and ion
exchange columns, which remove excess bromide from the color
developer solution, allow 99-percent reuse of the solution.

CASE STUDY 45

American Fotokemi, Inc., established a central processing
plant for recovering silver from films and spent fixer solutions
used in film processing. Conventional approaches to silver
recovery in the film processing industry have located canister
and electrolytic recovery equipment at each individual film
processing site. This decentralized approach results in less
than optimal recovery of silver due to sub-ideal process
monitoring and to the operating economics of small independent
recovery units. There is a complete loss of silver by small
processors who cannot economically justify recovery units.
American Fotokemi has developed chemical treatment processes that
increase silver recovery to nearly 100-percent -- an increase of
10 to 15 percent from that achieved by conventional decentralized
technology. Furthermore, other toxic metals, which would
otherwise be discharged as pollutants by the small, independent,
silver recovery operations, are also recovered. The high
efficiency of this centralized recovery method has increased the
dollar/silver return for everyone who has converted to it from
in-house canister and/or electroplating recovery.

Increasing the Rate at Which Industries and Institutions Implement Pollution Prevention and Waste Minimization Policies and Technologies

GOALS AND OBJECTIVES FOR WASTE MINIMIZATION THROUGHOUT SOCIETY

As our society moves beyond sole reliance on end-of-pipe pollution control approaches toward system-wide pollution prevention, waste minimization, and toxicity reduction procedures, an array of technical, economic, policy, and ecological factors must be addressed. To that end, the authors hope this book will help to stimulate interest in preventative approaches to waste management. We suggest the following goals and objectives as helpful for making this transition:

(1) Develop an awareness within the minds of our political, industrial, academic, and civic leaders of the ecological and economic advantages of managing our wastes through Pollution Prevention-Waste Minimization (PP-WM) approaches.

(2) Educate our citizens to view all wastes as resources.

(3) Reduce or eliminate the production of hazardous and non-hazardous wastes (resources).

(4) Reduce human and environmental risks due to exposure to toxic substances and hazardous wastes.

(5) Inform citizens in all sectors of society about successful waste reduction and waste elimination processes already being used by various industries worldwide. (The economic benefits, technological feasibility, and environmental assets of each procedure should be described in detail.)

(6) Develop, test, and utilize safe methods for converting wastes into useful products.

(7) Design new products and/or redesign old products to minimize the materials and energy used in their manufacture. The products should be designed to be energy- and materials-efficient throughout their life cycle; to release minimal quantities of wastes during

operation; and, to be easily and safely reusable, recyclable, or incineratable at the end of their useful life.

(8) Develop and implement processes for sharing concepts, approaches, and results, both within and across industrial groups. Each industrial category should also identify those problems for which there are currently no solutions. Governments should establish high priorities for researching these problems. These research results should be disseminated internationally.

It is our considered opinion that, by implementing these and other appropriate goals and objectives in the next decade, our nation's industrial firms will become more efficient and profitable. They will produce and release fewer pollutants, and the earth will be a safer place for all species.

A DISCUSSION OF IMPEDIMENTS TO THE IMPLEMENTATION OF POLLUTION PREVENTION-WASTE MINIMIZATION (PP-WM) APPROACHES

What impediments or barriers to the more extensive implementation of PP-WM currently exist? The following categories appear to us to be the most salient in resisting PP-WM approaches:

1) Regulatory officials. They have become accustomed to mandating end-of-pipe pollution controls, and are frequently unaware or unwilling to consider PP-WM approaches to attaining environmental quality goals.

2) Industry. The leaders and workers both have a tendency to resist change and to continue to insist that the regulators demand that they use pollution controls.

3) Academic circles. Professors, who were educated 10 to 30 years ago, learned pollution control philosophies and processes. Many of them have not heard about PP-WM approaches, or are not convinced that they are technically and economically sound. Consequently, they continue to teach new generations of students pollution control concepts, not PP-WM concepts.

4) Consulting engineers. For decades, they have been trained to design and have been designing pollution control installations. They are comfortable with these processes, and are usually unfamiliar and uncomfortable with PP-WM approaches.

5) Active environmental citizen organizations. Many of

their members are suspicious of government and industry. They frequently react to the concepts of PP-WM as "too good to be true". It is encouraging to note, however, that some environmentalists are actively promoting PP-WM approaches. They are doing so because they are convinced that governmental, industrial, and environmental interests can be brought together in non-competitive, mutually beneficial interactions, rather than in the typical, competitive, mutually harmful standoff.

Regulatory Barriers

Most of the environmental laws, rules, and regulations enacted during the last 15 years have focused on pollution control as the way to achieve the desired environmental quality goals. With a few exceptions, the laws have seldom focused on prevention. With this bias in the laws, regulators believe they have no choice but to enforce the laws through pollution control approaches. Additional regulatory barriers include:

1) Piecemeal coverage and inconsistent exemptions. For example, small-quantity generators are presently exempted from certain formal parts of RCRA accounting. Consequently, much of the hazardous residuals produced by small-quantity generators are known to be be mismanaged.

2) Fragmented jurisdictions.

3) Inconsistent enforcement.

All three of these barriers decrease the likelihood that PP-WM approaches will be implemented by companies large and small, in all parts of this country, or in all countries.

Additionally, the RCRA regulations themselves still make the recycling of potentially salvageable resources very difficult. Because of the morass of stipulations designed to prevent any unforeseen regulatory loopholes, hazardous by-product streams which are "sometimes discarded" can not be exempted from regulation, even if a buyer stands ready. This kind of non-discretionary and sweeping bureaucratic proclamation is seriously impeding the development of direct "waste" exchanges from generators to parties who can use the by-product as a resource.

The legislative and judicial branches of the government frequently act as major barriers to the implementation of PP-WM. This appears to be due to their commitment to uphold the letter of the environmental laws presently on the books.

The 1984 RCRA amendments do focus on waste reduction and, as such, are encouraging harbingers of a new regulatory era.

Economic Barriers

The first and foremost economic barrier to comprehensive implementation of PP-WM is the present availability of inappropriately low-cost land disposal for hazardous waste in pits, ponds, lagoons, landfills, and by deep-well injection. As a consequence, most industries have traditionally "disposed" of their wastes, rather than looking for ways to avoid producing them in the first place. Recently, as the public has become aware of the environmental risks associated with these methods, the price of disposing of the wastes has begun to rise rapidly. In many states and nations, these approaches are banned, or are being phased out in favor of other, more ecologically sound management approaches.

Informational Barriers

Because PP-WM concepts are comparatively new and innovative, many governmental, industrial, academic, and civic leaders have little or no information about them. The problems appear to be in the following areas:

1) Lack of sufficient theoretical and philosophical information about PP-WM concepts.

2) Lack of sufficient "proven" technical information about PP-WM approaches.

3) Lack of sufficient short- and long-run economic information about PP-WM concepts.

4) Lack of sufficient short- and long-run human health information about PP-WM approaches.

5) Lack of sufficient short- and long-run ecological information about the consequences of implementing PP-WM approaches.

Product Acceptability Barriers

Many citizens have an aversion to using products that contain reclaimed or recycled constituents. As a consequence, there are frequently barriers to the assembling, processing, and sale of recycled materials.

In addition, manufacturers inclined to use non- virgin materials are often confronted with an array of standards that virtually mandate the use of new materials.

In summation, an array of barriers is presently slowing down the implementation of PP-WM approaches in industries around the in industries around the world. Some of these barriers are more important than others, and will vary from country to country, and from industry to industry. The authors welcome your ideas and suggestions about any additional barriers to the implementation of PP-WM approaches.

PROPOSED STRATEGY FOR OVERCOMING THE BARRIERS AND ACCELERATING THE GLOBAL IMPLEMENTATION OF PP-WM APPROACHES

In relation to the goals and objectives outlined above, we wish to propose the following for your consideration as initial approaches to speed the global implementation of PP-WM.

First, and foremost in importance, is the need to focus on attitudinal barriers. One way to break through these barriers is to provide appropriately concise or technically detailed information for different audiences, be they citizens, educators, governmental officials, or industrial leaders. The information should document the PP-WM concepts and clarify how they differ from pollution control concepts.

The information should also explain the PP-WM technologies involved, the economic benefits to be gained, as well as the human health and environmental advantages. This information should be assembled on a case study basis for each PP-WM example. These case studies should be grouped by SICs and cross-indexed according to the problem addressed and the approach to solving it. This information could be made available in a loose-leaf format so that additional examples could be added at a later time. The information could also be made accessible by computer.

Such an assemblage of technical information, based on examples, and selected on a worldwide basis, would be a valuable educational tool for industrial leaders, government officials and university educators. The U.N. Low-Waste, Non-Waste Compendium (1982) is an excellent step in this direction. Other publications that also contain valuable information on pollution prevention approaches include Campbell (1982), Chazelon (1982), Huisingh (1982), and Royston (1979). (These publications are listed in full in the "Literature Cited" section.)

The appropriately written and targeted information should then be widely disseminated to:

- o Media
- o Government agencies (federal, state, and local)
- o Associations or groups of consulting engineers
- o Industrial trade associations
- o Educational institutions
- o Citizen environmental organizations
- o Business associations and chambers of commerce

Establish an International Journal of Waste-Reduction

In order to establish and maintain lines of communication between technologists and scientists developing new PP-WM solutions, an international journal on pollution prevention and waste minimization should be initated. It would provide a forum for presenting successful PP-WM case studies, results of waste utilization or reduction technologies, and for discussing facets of currently unsolved problems.

Establish an Annual Worldwide Symposium Series on Pollution Prevention-Waste Minimization Issues and Successes

Worldwide symposia should be planned by representatives from many nations and should include members and participants from government, industry, academia, and environmentally concerned citizen groups. Participants should also be drawn from these same diverse groups. The media should be invited to cover all symposia.

Since the cost of worldwide travel severely limits the number of people who can attend these symposia, we propose that global communications satellite networks be utilized to allow the symposia to be held simultaneously at several conference centers around the world.

This type of multi-location conference is especially effective if the conferees have received and read the informational materials well in advance of the meetings.

Incentives for Industry to Implement PP-WM Approaches

Tax breaks, loans, and the like may be needed to encourage more industries to hasten their incorporation of these approaches. The following incentives might be helpful:

1) Tax breaks for industries that install and utilize PP-WM approaches and demonstrate "real" reductions in the production or toxicity of wastes.

2) Tax-exempt industrial revenue bonds for underwriting industrial changeover to PP-WM approaches. 3) Encourage on-site reuse, incineration, and cogeneration of energy from wastes.

4) Support the use of waste exchanges.

5) Offer awards to industries that have made outstanding strides in reducing or eliminating waste production.

Disincentives for Industry to Continue Unsatisfactory Management of Residuals

In order to further hasten the conversion of industries to PP-WM approaches, the following actions should serve as disincentives to continuing unsafe practices:

o Ban the use of landfills and deep-well injection
o Enact and enforce strict liability laws
o Enact and enforce "waste-end" taxes
o Enact and enforce victim compensation laws
o Enact and enforce worker and citizen right-to-know laws

CONCLUSION

In summary, we have addressed an array of goals and objectives that we believe we should strive to achieve. There are barriers to rapidly achieving these goals, but we are confident that this combination of educational, research, legislative, and economic approaches will be a useful starting point for a global effort to foster the accelerated implementation of pollution prevention-waste minimization throughout all industries and throughout society.

In doing so, our air, water, land, and food will be cleaner and freer of toxic substances, and our people's health will be improved.

We welcome your comments, suggestions, and willingness to work toward the common goal of a safe, clean environment for the present and for the future.

LITERATURE CITED

1. Bringer, Robert, P. and Sara Zoss. "3P: An Environmental Protection Program that Works". Paper presented at the Low- and Non-Waste Technology Conference held in Tashkent, USSR, October 15-19, 1984.

2. Campbell, Monica, E. and William M. Glenn. Profit from Pollution Prevention. Published by the Pollution Probe Foundation, Toronto, Ontario, Canada. 404 pp. 1982.

3. Chazelon, Jean-Claude. Les Techniques Propres Dans L'Industrie Francaise. Published by La Societe Objective, 19 Rue Vaugelas, Paris, France. 1982.

4. Compendium on Low- and Non-Waste Technology. Produced by the Economic Commission for Europe and the United Nations Environmental Programme, Geneva, Switzerland. 1982.

5. Hazardous Waste Management: Recent Changes and Policy Alternatives. Congressional Budget Office. May, 1985.

6. Huisingh, Donald and Vicki Bailey, (eds.). Making Pollution Prevention Pay: Ecology with Economy as Policy. Published by Pergamon Press, Elmsford, N.Y. 156 pp. 1982.

7. Martin, Larry. A North Carolina Citizen's Guide to Best Management Practices for Hazardous Wastes. Nroth Carolina Citizen Action on Toxic Chemical Hazards, Siler City, N.C. 254 pp. 1984.

8. Protecting the Nation's Groundwater from Contamination. U.S. Congress Office of Technology Assessment. 1984.

9. Royston, Michael, G. Pollution Prevention Pays. Published by Pergamon Press, Elmsford, N.Y. 197 pp. 1979.

10. Superfund Strategy. Summary. U.S. Congress Office of Technology Assessment. March, 1985.

APPENDIX 1

The Questionnaire Used In Gathering Data For The Compendium,

Proven Profit from Pollution Prevention.

Dear Respondent:

The information requested in this questionnaire, along with similar types of information from other industrial firms, will be used to develop a compendium of case studies that will illustrate the benefits to industries and the environment from the utilization of pollution prevention approaches.

In filling out this form, please add attachments if more space is needed and, where applicable, please include tables, photographs, or flow diagrams. If there is sensitive information you would prefer not to divulge, we will honor whatever limits of confidentiality or anonymity you desire. If you have any questions about anything relating to this project or this questionnaire, please contact one of us.

Dr. Don Huisingh (919)-737-2470
Ms. Helene Hilger (704)-597-4175
Mr. Sven Thesen (919)-737-2479

1. Name of Firm:

2. Contact Person: 3. Title:

4. Address: (please include county)

5. Telephone:

6. Type of Industry:

7. Standard Industrial Classification:

8. Length of time the firm has been at its current location:

9. Approximate number of employees:

10. Type of Pollution Prevention Modification:

11. In a narrative and/or diagrammatic form, contrast the pollution prevention process with the previously used method.

12. Please complete the following table in order to compare the previous process to the pollution prevention process.

METHOD PREVIOUSLY USED/POLLUTION PREVENTION METHOD

Raw material
costs
Water:
costs & volumes

Energy costs

Wastes generated:
hazardous & non-
hazardous

Waste
management/
disposal costs

Pollution control
costs
Personnel
maintenance costs

Quality of
product

Productivity

Down-time

Environmental
impact

Health hazards

Other

13. What were the capital costs of implementing the pollution prevention process?

   a. Facilities:

   b. Equipment:

14. What are the annual net cost savings and the payback period due to the new process?
15. What time was required to plan and implement the modification?

16. How have the following factors encouraged, inhibited, or otherwise influenced the implementation of the pollution prevention process?

    a. Federal programs/regulations: (including influences from RCRA, NPDES, OSHA, the Air Quality Act, etc.)

    b. State environmental programs/regulations:

    c. Economic Factors:

d. Other Factors:

17. Please list the cosultants and equipment suppliers needed to install the pollution prevention process.

18. Please provide any overall comments or recommendations about this particular modification that might be useful to others in this industry or related industries.

19. Does your firm have an incentive awards program that encourages employees to submit suggestions of additional pollution prevention approaches? If so, please describe your incentive awards program.

20. Are there any problem areas associated with your pollution prevention process that need to be addressed at the State or Federal level?

21. Narrative Summary: Please summarize the general process(es), the description of the modification(s), and the rewards in terms of residues (wastes) not produced, raw materials not used, decreased environmental and human health risks, energy and money saved.

Please return this completed form to:

Dr. Don Huisingh
Division of University Studies
PO Box 7107
North Carolina State University
Raleigh, North Carolina 27695-7107
(919) 737-2470

Thank you for your valuable contribution to the POLLUTION PREVENTION COMPENDIUM.